CAMBRIDGE LIBRARY COLLECTION

Books of enduring scholarly value

Life Sciences

Until the nineteenth century, the various subjects now known as the life sciences were regarded either as arcane studies which had little impact on ordinary daily life, or as a genteel hobby for the leisured classes. The increasing academic rigour and systematisation brought to the study of botany, zoology and other disciplines, and their adoption in university curricula, are reflected in the books reissued in this series.

Memorials of Sir C.J.F. Bunbury

Sir Charles James Fox Bunbury (1809–86), the distinguished botanist and geologist, corresponded regularly with Lyell, Horner, Darwin and Hooker among others, and helped them in identifying botanical fossils. He was active in the scientific societies of his time, becoming a Fellow of the Royal Society in 1851. This nine-volume edition of his letters and diaries was published privately by his wife Frances Horner and her sister Katherine Lyell between 1890 and 1893. His copious journal and letters give an unparalleled view of the scientific and cultural society of Victorian England, and of the impact of Darwin's theories on his contemporaries. Volume 3 covers the years from 1848 to 1856. In 1848 Bunbury published a well-received account of South Africa, and particularly its natural history: he had written it after visiting Sir George Napier, his step-mother's father, then governor general. Another notable event was a meeting in Berlin with Alexander von Humboldt.

T0174449

Cambridge University Press has long been a pioneer in the reissuing of out-of-print titles from its own backlist, producing digital reprints of books that are still sought after by scholars and students but could not be reprinted economically using traditional technology. The Cambridge Library Collection extends this activity to a wider range of books which are still of importance to researchers and professionals, either for the source material they contain, or as landmarks in the history of their academic discipline.

Drawing from the world-renowned collections in the Cambridge University Library, and guided by the advice of experts in each subject area, Cambridge University Press is using state-of-the-art scanning machines in its own Printing House to capture the content of each book selected for inclusion. The files are processed to give a consistently clear, crisp image, and the books finished to the high quality standard for which the Press is recognised around the world. The latest print-on-demand technology ensures that the books will remain available indefinitely, and that orders for single or multiple copies can quickly be supplied.

The Cambridge Library Collection will bring back to life books of enduring scholarly value (including out-of-copyright works originally issued by other publishers) across a wide range of disciplines in the humanities and social sciences and in science and technology.

Memorials of
Sir C.J.F. Bunbury

VOLUME 3: MIDDLE LIFE PART 2

EDITED BY
FRANCES HORNER BUNBURY
AND KATHARINE HORNER LYELL

CAMBRIDGE
UNIVERSITY PRESS

CAMBRIDGE UNIVERSITY PRESS

Cambridge, New York, Melbourne, Madrid, Cape Town,
Singapore, São Paolo, Delhi, Tokyo, Mexico City

Published in the United States of America by Cambridge University Press, New York

www.cambridge.org
Information on this title: www.cambridge.org/9781108041140

© in this compilation Cambridge University Press 2012

This edition first published 1891
This digitally printed version 2012

ISBN 978-1-108-04114-0 Paperback

MEMORIALS

OF

Sir C. J. F. Bunbury, Bart.

EDITED BY· HIS WIFE.

THE SCIENTIFIC PARTS OF THE WORK REVISED BY
HER SISTER, MRS. LYELL.

MIDDLE LIFE.

VOL. II.

MILDENHALL :
PRINTED BY S. R. SIMPSON, MILL STREET.

———

MDCCCXCI.

LETTERS.

CHAPTER IV.

My Dear Katharine,

I am afraid you will have lost all patience 1848. at my delay in answering your questions about the Mosses; but I have really had a great deal to do, and very little time to spare, and having been for a good while rather out of the habit of examining Mosses, I couldn't determine them at first sight; some indeed I do not feel sure of now; but I will put at the end of this letter, such information as I can give you concerning them. I am delighted to hear that you have been botanizing with such ardour and perseverance, and that your zeal has been rewarded by such a discovery as that of Buxbaumia Aphylla, which is indeed a prize. I am half inclined to envy you, for I don't think I ever found anything so rare. I hope the discovery has been communicated to Sir William Hooker. I collected a few interesting

3 B

1848. Mosses and Lichens, (principally indeed the latter)
on the Alps, and shall have great pleasure in
sharing them with you, when you will pay us a visit.
We reckon very much upon the pleasure of having
you and your Husband here with us, and you must
give us a good long visit, that we may get well
acquainted with him. Your botanical pursuits will
be a great resource to you in India, and I hope you
will be able to go to the Cape, where you will find a
great deal to please you, especially if you are there
in the right season, which is from June to
November.

I have been looking at a collection of Mosses
from Kinnordy, which I think were sent me by
some of the Miss Lyells, just before we left England,
when I had not time to look them over. Pray give
my best thanks to the donors. I find several things
which I am very glad to have, especially Neckera
Curtipendula in fructification, which is rare; and
all the specimens are in very fine condition.

Charles and Mary have just sent us your list of
forty-six Mosses from Kinnordy, which I was very
glad to see, as I confess I have been a little
sceptical. Kinnordy may well be proud of such a
list.

I had no idea that you had found the fructification
of Hypnum Dendroides and Dicranium undulatum,
which are excessively rare in that state. But I
infer, as it is absent from your list, that you have
not yet discovered Hypnum Crista castrensis in
fruit.

It is abominably cold here and everything out of

doors looks gloomy and untidy, but it is a great 1848. comfort to be at home again, and to find our house and goods in excellent order and condition, and our favourite plants in the garden thriving.

I enjoy being at work again among my books and collections, but for the present I have so much to do to get a paper ready for the Geological Society against the 29th, that I have little time for arranging my Swiss and Italian specimens.

I have plenty of occupation before me for the winter, as there are fourteen large boxes of dried plants, from Brazil and North America, ranged in goodly array in the long gallery, and waiting for examination. They belonged to my uncle Mr. Fox. I have not had time yet to open a single one of them.

We shall be coming up to town about the 27th. I am afraid you will not be there so soon, but I wish you may arrive before we come back to Suffolk. We have so much to do here, that we shall be rather unwilling to stay long in London at this time, we shall be returning to it in the spring. We both of us keep well hitherto in spite of the cold, and although the fever has not ceased in the parish, I am not afraid of it.

Pray give my love to your Husband, whom I hope to know much better than I do yet; and remember me very kindly to Mr. Lyell and the Miss Lyells.

Believe me ever your affectionate Brother,

C. J. F. BUNBURY.

Mildenhall,
February 5th. 1849.

My Dear Mary,

1849.   I thank you very much for your kind note
and your good wishes. I feel all the weight and
gravity of *middle age* descending on me, now that I
have turned the awkward corner of forty, though
your Husband, who turned that corner a good while
before me, seems as brisk and juvenile as if he
had it still in prospect,   I am very sorry to hear
that his book proves a longer undertaking than he
had contemplated, and that we must not reckon
confidently on reading it this summer. I hope he
does not mean to overload it with politics or
statistics.

I like Henry Lyell more and more as I get better
acquainted with him : his manners are very pleasing,
and he seems to me very right-minded, very clear
and just in his views, and very desirous of information, as well as most thoroughly gentleman-like
in every respect ; nor do I find his deafness quite so
great an impediment to conversation as I had
expected.   I have had some good talks with him
about India.

Katharine and Joanna's society is a delightful
treat to me, and (what you will perhaps think
wonderful) I listen with *real* pleasure to Joanna's
singing. With Katharine I have had many satisfactory confabulations about botany, and am
delighted with her zeal and earnestness in the
pursuit. It is a great pleasure to me to show her

my collections, and I have been able to give 1849.
her several good Mosses from Brazil and other
countries.

Ever your very affectionate,

C. J. F. BUNBURY.

[I cannot find any written materials between
February and September. I believe we went to
London for a short time, but we spent most of our
time at Mildenhall, with visits from our relations on
both sides. I especially remember a visit from
Lady Napier and Mrs. John Napier and her little
girl Sarah, and also one from Mrs. William Napier
and her little girl Cecilia. Towards the end of
August we went to Scotland. We first spent some
days in Edinburgh, and then went to Kinnordy,
a place belonging to Mr. Lyell, Sir Charles Lyell's
Father, where I had a dangerous illness.]

F. J. BUNBURY.

To Lady Bunbury.

Kinnordy,
September 14th, 1849.

My dear Emily,

    I am very happy to hear so good an
account of you all, and that the excursion to
Scotland has answered so well in regard to health
as well as pleasure. I feel your kindness in what

1849. you say about considering my taste in the matter of the Arboretum, &c., and I have no doubt that I shall approve of your alterations, as I almost always do; for I have a high opinion of your taste and I know that you have as profound a respect for handsome trees and shrubs as I have. Many of my acquaintances (and my Brothers among the number) are possessed with a Dendrophobia, which would make them very unsafe advisers; I am sometimes afraid of Fanny being infected by it. There are some very fine trees here, though not I think any rare or curious ones :—two noble Silver Firs, some of the finest Limes I ever saw, Willows and common Maples of extraordinary size, and a Bird-cherry far surpassing all my ideas of the size to which that tree would attain. Two Larches, which Charles Lyell tells me were uncommonly fine trees, were destroyed by lightning some years ago.

I was very sorry for the death of the good Bishop of Norwich.* His loss will be much felt in various ways. It seems there is a rumour that Whewell is likely to be his successor.

I grieve for the fall of the Hungarians and for the formidable ascendancy which it will give to Russia and to military despotism. How the state of affairs and the proper object of one's apprehensions have changed within the last half-year! Not very long ago, order and civilization seemed everywhere in danger from the violence and lawlessness of democracy; now, all Europe seems in danger of being *cossacked*. Well, I think all this makes it clear that

---

* Bishop Stanley.

no nation has any chance of preserving its freedom 1849.
which cannot maintain it by the strong hand, and
that as mankind are now constituted, the main-
tenance of a military spirit, and of military skill and
science, and of a good strong army too, is an
absolute duty of every people that does not wish to
be enslaved.   When Mr. Cobden can persuade the
Czar to disarm, we may allow him to be worth
listening to; till then, I own that I prefer the old
Cromwellian precept :—"put your trust in heaven,
and *keep your powder dry.*"

Much love from both of us to you and my Father
and Cecilia.

> Believe me
> Ever your very affectionate Son,
> C. J. F. BUNBURY.

From his Father.

Barton,
September 27th, 1849

My Dear Charles,

I do indeed feel for you, and pray for you
and for your poor Wife.   It is a terrible trial; and
we cannot disguise from ourselves the dangerous
condition of dear Fanny's health.

The season is so far advanced, we are drawing so
near to cold weather, that I trust you will continue
to move Southward (though by slow journeys),
when Fanny is able to bear travelling; and that
you will pursue your route to Hastings, or to

1849. the Undercliff, rather than tarry long at Edinburgh.

It is a comfort for us to think that Mr. and Mrs.
Horner are with you, both for Fanny's sake and
your own.—Assure them, with my kind remembrance that I partake sincerely in their present
anxiety.

The Indian mail did not bring me a line from
Henry (nor did the last) but Sir Charles Napier tells
Emily that your brother is very well, and that he is
a very clever fellow and a very good fellow. May
God bless you my dear boy and preserve you and
yours.

I am ever most affectionately yours,

H. E. B.

3 Rutland Street, Edinburgh,

Friday, November 2nd, 1849.

My Dear Emily,

I am sorry to say that Fanny has had a
slight relapse,—comparatively very slight, but still
it makes one uncomfortable.

Last Sunday—a most lovely day. I had a very
pleasant walk to Craigcrook, and saw Lord Jeffrey
and Mrs. Jeffrey, and Lady Bell, who is on a
visit there.

The other day we had a visit from Woronzow
Greig,* who was in Edinburgh for a few days, he
told me that Mrs. Somerville is quite well.

* Son of Mrs. Somerville.

Pray tell Cecilia that I am delighted at her 1849. taking up a fancy for natural history,—very glad too that she likes the "Seaside book," which I think charming. Is she aware that the author is our Cape friend, Harvey ?

I bought the "Friends in Council"* on your recommendation, and have been reading several of the essays, with great pleasure and satisfaction. Those on Reading, on Criticism, and on the Art of Living, I quite agree with you in thinking excellent : that on Education (in the first volume) also appears to me to be full of good and just thoughts, and above all I cordially approve and admire what he says against the practice of holding up "what will people think?" as a bugbear to children.

The Essay on Government is less vigorous, but I am delighted with the remark (in page 174) on the tendency of the day to "vulgarity of thinking in government." Humeism† and Cobdenism to wit! I rejoice to meet with a writer who, without any prejudice against what is modern, merely because it is modern, or any superstitious veneration for the wisdom of our ancestors, ventures to see, and to say that the present age is not the best of all possible ages, and points out the false direction (in some respects) of our recent civilization. I like the setting of the essays, the dialogue appears to me more spirited and natural than is usual in modern writing, and the characters are well kept up. I like the author too for his evident love of dogs.

* By Sir Arthur Helps.   † Mr. Hume was born 1777, died 1837.

1849.   I had my Father's narrative sent me by the post*
and have read it with very great pleasure, it appears
to me singularly clear, simple, and animated.
Fanny too is delighted with it, and she sketched a
plan of the scene of action from his descriptions. I
have also been engaged with the second volume
of " Cosmos," and have been particularly interested
by the chapter on the Oceanic discoveries of
Columbus and his contemporaries. I would recom-
mend this second volume to you—by-the-bye I am
not sure whether you have not read it : it is
much easier to understand than the first, and of
more general interest.

Our friends at Craigcrook the other day advised
me to spend the winter at Edinburgh; and Lady Bell
declared that the winter climate of this place (before
February) was the mildest she knew.

I am sorry that Edward persists in his intention of going so far as Athens this winter.
Greig seemed to have been struck by his reduced
and enfeebled appearance.

Poor Mr. Lyell seems to be really dying, and
unfortunately amidst great suffering : his medical
attendant has given him over, and his own family
can but wish that he may soon be released from
pain.   Charles has been sent for, by electric

---

* I think it was in the course of the year 1848, that my Father was induced
principally by the advice of his wife, to set about the work of putting into a
connected and durable form, the many interesting facts with which his
memory was stored, relating to the great war with Revolutionary France.
He took up first the Campaign of 1799, in North Holland, and having connected and arranged his personal reminiscences, by the help of published
documents, he had his ' Narrative ' printed in the form of an 8vo. pamphlet
and distributed it to his friends, in the latter part of the year 1849.

Memoir of Sir H. E. Bunbury, by his Son. Page 192.

telegraph, and is by this time at Kinnordy. The 1849. poor old man seems to retain his fine mind and generous unselfish character to the last.

Pray give my love to my Father and to Cecilia, and believe me,

Your very affectionate Step-son,
C. J. F. BUNBURY.

To Lady Bunbury.

3, Rutland Street, Edinburgh,
Friday, November, 15th, 1848.

My Dear Emily,

I owe you many thanks for your letter of the 7th. I quite agree with you as to the advantages of Edinburgh ; it has certainly a far greater variety of resources than Brighton, or any other watering place, and though I have not the means of going on with those precise pursuits in which I was engaged at home, I am not afraid of time hanging heavy on my hands. I have met with much civility and hospitality, especially from Lord Murray, at whose house I passed two very pleasant evenings, and by whose assistance I have established an acquaintance with Mr. Macnab of the Botanic Gardens. I have an admission to the Advocate's Library, which is an excellent one ; and besides that the booksellers' shops are numerous and well supplied, so I have no fear of reading materials failing me. But after having so long enjoyed the use of poor Mr. Lyell's library, where the abundance of books as well as the

1849. peculiar circumstances of our stay at Kinnordy, rendered my studies rather desultory, I have found it a good plan to put myself on a short allowance of books for a time.

Having finished the " Cosmos," I am now deep in " Bacon's Advancement of Learning," with which I am delighted, but which is good stiff reading, and requires close attention. I occasionally read a canto of Dante with Fanny, who has studied him so much that she is as good as a commentator, and is very anxious to inspire me with a proper admiration of the old Florentine, a result not yet completely achieved. I have also read over again, with fresh pleasure, an old favourite of mine—" Gibbon's Life," and Fanny, to whom it was new, has been quite captivated by it.

Mrs. Horner and Leonora set off to-morrow for their home, having been assured by the Doctor that they may leave Fanny without anxiety. We shall miss them, but it is perfectly reasonable that they should wish to be at home again after so long an absence.

You will have heard of Mr. Lyell's death, which happened on the 8th ; to himself it was a happy release from long and severe suffering ; but he was a man to be regretted ; and his daughters especially, who had for a long time devoted themselves to him in a most exemplary manner, will long feel a sad blank. The accounts we have heard of his last day are very touching. To the last he retained the powers of his refined and noble mind, quite unclouded, was full of thought and consideration for

all his friends, and had a kind word for every one 1849.
around him.

As for weather, we have had a week or more of
warm rain, muggy, muddy and disagreeable ; but it
has now again become clear and cold, and I hope
to enjoy some walks.

Pray give my love to my Father and Cecilia, and
believe me

Ever your affectionate Step-son,

C. J. F. BUNBURY.

3, Rutland Street,
Nov. 18th, 1849.

My dear Mr. Horner,

I saw Mrs. Horner and Leonora safely
deposited in a York carriage of the express train
yesterday morning.

Poor Fanny was very unhappy at parting from
her kind Mother and Sister, who had devoted themselves with such indefatigable tenderness to her
service for so long a time.

We are most fortunate in our lodgings ; and
between the books we have with us, the resources of
the booksellers' shops (which seem remarkably well
supplied), the Advocates' Library, to which I have
an admission, the Museum and Botanic Garden,
and the beautiful country—to say nothing of society
—I have no fear of time hanging heavy on my
hands. I have already met with much civility and
hospitality, especially from my Father's old friend,

1849. Lord Murray, and have passed two evenings very
pleasantly at his house (in spite of the music),
and another also quite to my taste at Sir James
Gibson Craig's. The weather has been for the most
part so unfavourable, that I have made few excursions. I have however rambled over Arthur's
Seat and Salisbury Crags in various directions,
hammer in hand, studying them with very great
interest, collecting plenty of specimens and comparing my observations with Nicol's and Maclaren's
descriptions. It would hardly be possible to find
a more interesting exhibition of igneous rocks in
connexion with the stratified. I have been rather
disappointed with the museum : it strikes me as
much inferior, I will not say to those of the great
capitals, but to those of Turin and Strasbourg ;
it seems indeed to be chiefly rich in birds, which
I do not understand ; but the collection of minerals,
though pretty extensive, is by no means splendid,
and in fossils it seems remarkably poor. Of fossil
plants there is not one single specimen ; and
Dr. Flemming tells me that there is nothing to
be called a collection of them in this country.
Nobody in Scotland seems to have attended to
them, Professor Jameson, I understand, sets no
value on fossils and pays no attention to them.

The Botanic Garden is a noble collection : many
of the specimens of Palms and other tropical trees
are the finest of the respective kinds in Europe ;
but Mr. Macnab sighs over the want of funds for
building hothouses of suitable dimensions, and
envies the resources of Kew.

Mr. Lyell's death, though a happy release for 1849.
himself, must be a subject of regret to all who knew
him, as well as a deep affliction to his own family.
It is consoling and ennobling to see a fine mind
retaining its powers to the last, and triumphing over
pain and decay.

Fanny sends much love.

Ever your affectionate Son-in-law,

C. J. F. Bunbury.

3, Rutland Street,
November 30th, 1849.

My Dear Father,

I return to you Henry's letter, with many
thanks for it and for two notes which I have received
from you. I have little to tell you, as we lead very
quiet uniform lives. Fanny is pretty well, and
going on, I trust, satisfactorily. The prevailing
bad weather and some other circumstances have
kept her much confined to the house lately, but
she reads a great deal and with great enjoyment,
and her seclusion has been cheered by several
visits from her old friends here; and Mrs. Jeffrey
kindly supplies her with flowers, which she delights
in. I gave your message to Lord Murray, who
says he should be delighted to have your Narrative.
If you have a copy to spare, let me offer a
petition on behalf of Harry Lyell who is very
anxious to have it, and to whom as a soldier, it may
be instructive as well as interesting; he is going
southward, and will stay some time (a month I

1849. think) with the Horner's at Rivermede, and the rest
of the winter in London.

The Charles Lyells are again settled in Harley
Street.

Lord Murray is most kind and hospitable; 1 was
at a party at his house last night—too musical for
my taste.   And I dine with him again to-day.

Fanny is reading Middleton's Cicero, and I am
much interested in a new work of Humboldt's—
"Aspects of Nature"—which is a kind of recapi-
tulation or summary of what he has observed in his
travels, with an immense quantity of illustrative
matter from other sources.   Both our graver studies
have been interrupted for a time by the first
volume of "The Caxtons,"* which delights us;
have you read it?

Fanny sends her love.

Believe me,

Your very affectionate Son,

C. J. F. BUNBURY.

----

3 Rutland Street,
December 3rd, 1849.

My Dear Mr. Horner,

My Dear wife is going on comfortably, and
I am happy to say that neither her health nor her
spirits appeared to have suffered as yet from confinement to the house.   Her cheerfulness is really
admirable; her mind is always active, and as she

* A novel by Lord Lytton.

has here neither old women, nor school, nor garden to take up her time, she gets through a great deal of reading.

The newspapers speak of "premature winter," in England; here we have nothing of the sort, but almost continual wet and fog, with a mild relaxing temperature,—rather trying to the health and spirits and very unfavourable to the enjoyment of the scenery. One day last week was frosty, delightfully clear and bright, and bracing, and allowed me to enjoy a good ramble round Arthur's Seat; but the very next day came again an odious thaw.

I have made repeated visits to the Botanic Garden, which is a great resource to me. The magnificent Palm-house not only delights my eyes and my imagination, recalling and freshening up in my mind the pictures of the glorious tropical vegetation, but it also affords me important opportunities for botanical studies, and has enabled me to become much better acquainted with the structure and characters of Palms than I was before.

Through Lord Murray's kindness, I have become acquainted with Dr. Balfour, the professor of botany, whom I find very pleasant, communicative and obliging. I shall be much obliged to you for introductions to Mr. Milne, Professor Forbes and Hugh Miller, though I have not yet used the letters which you did give me, to Professor Jamieson and Mr. Maclaren. I met Professor Forbes at the Deanery at Ely; after the Cambridge scientific meeting, but a note from you would probably be useful in helping me to renew the acquaintance.

1849.   I dined at Lord Murray's on Friday—a very
pleasant party—Mr. Lefevre,* Mr. Rutherford, and
Mr. Thomas Thompson. Lord Cockburn is very
kind in coming often to see Fanny, and chat with
her.

(December 4th). Katharine and Harry spent
yesterday evening with us, and we had a glimpse of
them again this morning, amidst the sleet and snow,
for they brought winter with them.

I am heartily glad to hear of the well-deserved
token of merit which the Royal Society has voted to
Murchison ; and I have been still more rejoiced to
learn that the Linnean Society is about to honour
itself by unanimously electing Robert Brown its
President.

Give my love to Mrs. Horner and all your family
party.

Your affectionate Son-in-law,

C. J. F. BUNBURY.

----

3. Rutland Street,

Dec. 21st, 1849.

My Dear Mr. Horner,

I was much obliged to you for the notes
of introduction you sent me. Professor Forbes
received me most cordially, remembered quite well
our meeting at Ely, introduced me to the Royal
Society ; and seems anxious to be of use to me.

We have settled to go together some day to see
Hugh Miller and his collection. But just now I am

----

* Afterwards Sir John Lefevre, Poor Law Commissioner.

a prisoner; for after I had been teased for about 1849.
ten days by a troublesome little cough, which would
not go away, Fanny insisted on calling in Mr.
Newbigging, who has prescribed, besides other
things, confinement to the house for the present.
I am now better, and am not unwilling to stay at
home, having much occupation both in reading and
writing.

The letters from Rivermede are very pleasant and
comfortable to us. Joanna's letters are so full of
the Orgue expressif, that it is clear she is under
the influence of an *organic affection*.

I am going on in a way with my work on fossil
plants : not that I can complete any part of it here,
but I am able to sketch out the different parts, and
arrange them in a preliminary way, drawing up
from the books I have here and from memory,
an outline or a sort of *syllabus* (as I recollect the
lecturers at Cambridge used to call it), to be filled
up when I have more materials at hand. I have
lately read the " Old Red Sandstone," for the
second time : it is a very fascinating and very clever
book, but I think there is a little tendency to *romance*
here and there, especially in what he says of the
Carboniferous period; at least, the author's lively
imagination inclines him to adopt the more imagi-
native rather than the more sober theories of
geology. I am now reading for the second time,
Lyell's "Second Visit;" and for the first time,
Middleton's " Cicero ;" both with great interest.
Lord Cockburn comes often to see us, and is very
agreeable. We had likewise an extremely agreeable

1849. visit the other morning from Sir Frederick Adam. Our time is well filled, and though we cannot but sometimes feel a longing for our home, and often wish that we could see those who are most dear to us, we have reason on the whole to be satisfied with our situation.

I am sure I shall be well content with this winter in Edinburgh, if it restores the health of my dear Wife, whom I learn every day to value more and more.

To-day is Mrs. Horner's birthday. Pray give her my love, and say that I heartily wish her many happy birthdays to come. Much love to all your family party.

Ever your affectionate Son-in-law,
C. J. F. BUNBURY.

3, Rutland Street,
December 23rd. 1849.

My Dear Father,

I thank you very much for your last letter, which I have read with very great interst. I heartily concur in your dislike of the Manchester school of politicians, and of all who would reduce Government to a mere question of pecuniary profit and loss. I regard our Colonial possessions as both honourable and useful to us, if properly managed ; though as you say, we kept too many, and not a very judicious selection in some cases I think. But I do not think we can or ought to look upon our Colonies as strictly permanent possessions. It seems to be in the nature of things that a Colony if well managed, 1849. should in process of time grow up to be so wealthy, strong and intelligent, that the people will be both able to take care of themselves, and naturally and reasonably anxious to do so; and then surely it would be as unwise for the mother country to endeavour to retain them in subjection, as for a bird to keep its young ones in the nest after they are fully fledged. I speak of Colonies properly so called, and not of military posts such as Malta and St. Helena. It seems to me therefore, that our object should be, not to govern our Colonies for our own exclusive benefit, but so as to give them the best training for future self-government; to look upon an ultimate separation as unavoidable, or rather as belonging to the laws of nature; and so to prepare for it, that it may at last take place amicably and not as the result of a struggle, which would leave a lasting soreness on both sides. The American war is a lesson which surely ought not to be forgotten. I do not mean to say that any of our Colonies are yet properly ripe for independence; though I am not at all sure about Canada; but I think we must expect them all to drop off in process of time, and that if we manage wisely, the separation may take place on terms advantageous to both parties, and we may find them useful allies and friends after they have ceased to be subjects.

Then comes the question whether our Colonial Government has been and is wisely adapted to this end; and I am afraid it certainly is not. The supreme direction of such a Colonial Empire as

1849. ours, seems almost too much for any man; but I think we have been particularly unfortunate in such Colonial Secretaries as Lord Stanley and Lord Grey. I do think the West Indian Colonists have been very ill-used, and their determination to reduce the official salaries, when their means of paying them have been so much reduced by the measures of our Government, seems but reasonable. I am not sure that the English settlers at the Cape did not sympathize with the irritation of their Dutch neighbours *before* this convict question; but *this* certainly has had the effect of uniting men and parties in the Colony that were previously the bitterest opponents.

It is something quite new to find Mr. Fairbairn of the *Commercial Advertiser*, at the head of a party including the principal Dutch families with whom he used to be almost as great a *favourite* as the missionaries or Sir A. Stockenstrom. After all, the Cape people seem to me unreasonable even in the ground of their opposition, and most culpable in the extent to which they have carried it. But as you say, to push this measure at first so hastily as Lord Grey has done, and then to retract it on meeting with seditious opposition, is likely to have a very bad effect. Sir Harry Smith appears to have at first sympathized with the Colonists and treated them with rather too much indulgence; in return for which they have tried to starve him and his troops. I lament, with you, the want of a strong Government. I am afraid there is no man fit to grapple with the existing difficulties, except Sir Robert Peel,

and has he any party ? As to Ireland, the sub- 1849.
sidence (for the time) of the anti-English agitation—
for even Mr. Duffy professes to be content to devote
his attention to social questions—appears to leave
an unusually fair opening for a bold attempt to
remedy the enormous and inveterate social evils of
that country ; but only a strong and very bold
Ministry could prosper in such an undertaking.

In reference to England, I think you have
omitted to notice one serious reason for uneasiness :
the excessive mutual rancour and bitterness between
the agriculturists and the manufacturers ; a rancour
which seems to grow more and more virulent on
both sides, and I really do not know which party is
the more unreasonable. I am afraid the coming
session will be both a stormy and an unprofitable one.

We had a very agreeable visit the other day from
Sir Frederick Adam. He tells us that he is in
correspondence with you about your Narrative, and
he is very desirous that you should write the history
of the brilliant campaign in Egypt, which he says
has never yet been well written. I heartily hope
you will do so, and will complete your " Memoirs
of the War " down to the Outbreak in Spain ;
the approbation with which the fragment you
have already printed has been received by all
who have seen it, is surely an encouragement to go
on. In the case of the Egyptian campaign, indeed,
you would not have the same advantages of personal
knowledge, but you knew most of those who were
engaged in it, and especially Sir John Moore, and I
suppose must often have heard it talked over.

1849.   Sir Frederick said that he thought Sir Ralph
Anstruther might have, among his father's papers,
some information relating to that singular plot
among the superior Officers of the army in Egypt
against Lord Hutchinson, of which there seems to be
no published account, and if so he is sure he can
obtain it for you.

Fanny lent your Narrative to her old friend Lord
Cockburn, and he too thought it excellent. He was
particularly struck with the truth of your account of
Sir David Dundas, who was a friend, and I think a
connection of his ; he says it is the man himself.

I have been confined to the house for nearly a
week, by order of Dr. Newbigging, whom Fanny
insisted on calling in, as my cough was so obstinate.
Fanny and I spend most of our time in reading,
but this cough of mine has vexatiously put an end to
my reading aloud to her, as I used to do in the
evenings.

I am interested in Middleton's "Cicero," the picture of a wealthy, powerful, ambitious, free community
in its latter days, is striking, and sometimes makes
one think of our own country ; but I find more
points of contrast than of analogy ; our chief evils
and dangers are materially different from those
under which the Roman Republic sunk. The
enormous profligacy and corruption of the judges
in the latter days of Rome, is what most strikes me.
A Republic is indeed ripe for destruction when the
judicial body is so utterly corrupt and venal as it
was in Cicero's time. There is no evil I think,
from which we are more secure than from this. The

wealth and prodigality of the principal citizens in 1849. Cicero's days, are something perfectly astounding. But I must come to a conclusion.

Pray give my love and Fanny's to Emily and Cecilia, and believe me

Your very affectionate Son,

C. J. F. BUNBURY.

P.S.—I have been reading with great pleasure and interest " Carlyle's Life of John Sterling." Pray read it if it comes in your way. It has all Carlyle's best qualities with fewer than usual of his faults; and without any remarkable incidents, without any striking facts, it is really one of the most interesting and attractive biographies that I have met with. The style seldom deviates into those rugged uncouth barbarisms that Carlyle is prone to, and is often quite admirable, and in his opinions and sentiments I find more that I can sympathize with than in any other of his writings that I know. Especially, the chapter on Coleridge is capital. I met this Mr. Sterling at Sir C. Lemon's, in the Autumn of 1841.

———

NOTE. — Extract from "Life of John Sterling," by Thomas Carlyle. Page 276. August 29th, 1841.—" I returned yesterday from Carclew, Sir C. Lemon's fine place. Met there (among others) Mr. Bunbury, eldest Son of Sir Henry Bunbury, a man of much cultivation and strong talents." [I quote this from John Sterling's diary.—

F. J. B].

3, Rutland Street,
December 31st.
(Last day of old 49).

My Dear Emily,

1849.

Very many thanks for your interesting letter, and for sending me Edward's improved report of himself, which gave both Fanny and me very great pleasure, as we had been much grieved to hear of his illness.

I am glad to find that you and I are really very much of one mind with respect to the proceedings in Cephalonia, and that our apparent difference of opinion arose merely from the difference of the statements of fact on which we proceeded. It is by no means my province to answer for Sir Henry Ward's veracity; I do not know him; and if the account you first heard was correct to the full extent, in its obvious and literal meaning, I quite agree with you as to the atrocity of such deeds. Either you or I have entirely mistaken the meaning of Milverton's* saying, which I quoted. I certainly do not understand him to mean that defeated insurgents should be treated with severity, much less cruelty after resistance has ceased; but simply that there should be no dallying or trifling with insurrection; and that if physical force be necessary it should be extended at once and vigorously, so as if possible to prevent the rebellion from coming to a head. How often in '48 and the beginning of this year has a mere *émeute*, which at first might have been put down with little bloodshed, been

---

* Quoted from " Friends in Council," by Arthur Helps.

allowed to grow up into a formidable and even 1849. a successful insurrection. The way that the London Chartists were dealt with in April, 1848, seems to me a model of the right way of dealing with such matters as, on the other hand, the conduct of the Grand Duke of Tuscany appears a model of the wrong way of dealing with them. But I own I should be much puzzled by a *passive* resistance such as that of the Cape people. I admit there is a great deal of truth in your qualifications of the maxim respecting the opinions of others. It is a very nice and delicate question *how far* public opinion ought to be allowed to have weight; but I have not room to enter upon it now. I hope my Father is well and that he received the letter I wrote to him in answer to his on politics.

You say nothing about leaving Abergwynant, and I daresay you find it pleasanter at this season than Suffolk. I wish we were in the South. I wish you all a "a Happy New Year," and sincerely pray that it may be a year of health and prosperity and comfort to all of you. I am reading Dr. Hoffmeister's "Travels in India," and find a good deal of entertainment in them. He was the medical attendant of Prince Waldemar of Prussia, and was killed while riding by the side of Lord Hardinge at Ferozeshah.

Believe me ever,

Your affectionate Step-son,

C. J. F. Bunbury.

1850.   Edinburgh,
                                    January 10th, 1850.

My Dear Lyell,

I have lately read over again your "Second
Visit," and liked it fully as much as at first: it has
not, however, tended at all to make me a convert to
democracy,—quite the contrary; but I will not
enter upon politics just now. Twice, and only
twice in the course of the book, I felt tempted to
exclaim how I should like to be there!—first, in
reading of the White Mountains, and again, when
you were at Mr. Hamilton Cooper's in Georgia.

I am struck with what you mention of the severe
cold experienced in Georgia, even on the sea
coast, Savannah is nearer to the equator than the
Cape of Good Hope, where frosts are quite unknown,
except at a considerable elevation. Yet the evergreen Oak of Carolina and Georgia—the Quercus
virens—will not bear even our ordinary English
winters; probably because of the damp which
accompanies our cold.

Have you any information about California in
your American letters? Is it true that the gold
diggers have made a constitution for themselves,
and excluded slavery? By the account in the
Times, several of the provisions of their proposed
constitution seemed to be exceedingly wise. It is
certainly a marvellous thing that such a promiscuous
and (as it seemed at first) lawless horde, drawn
together by no other motive than the thirst of gold,
should in so short a time become sensible of the
want of law and government, and set themselves so

vigorously and wisely to work to remedy it. 1850.
Truly there is a wonderful energy in these same Anglo
Americans. But what a strange mess the Government and Congress seemed to be in at Washington ;
—fairly brought to a dead lock. Is it true that
the *Mormons* are going to found a State by
themselves ?

I suppose you saw in the Athenæum that Miss
Martineau's Eastern Book has been solemnly *burned*,
as profane and impious, at Burton upon Trent ?
Lucky that the inquisitors could not get hold of the
authoress herself. As there is some burning
matter in your " Second Visit ;" I hope you will also
keep clear of Burton upon Trent.

Much love to Mary.

Ever your affectionate friend,

C. J. F. BUNBURY.

3, Rutland Street,

January 31st, 1850.

My Dear Mr. Horner,

I thank you much for your kind letter of
the 28th. I knew you would be grieved at the
death of Lord Jeffrey; it makes, as might be expected, a great sensation here, and it is evident
that he is very generally regretted; the party
feeling once associated with his name seems to have
long since subsided, and I understand that the
regret is shared by persons of all political opinions.
His illness was very short, and he seems hardly to
have been considered in serious danger till the very

1850. day of his death; the evening before (Friday) I was dining at Lord Murray's, and heard that Lord Jeffrey was better, and no one appeared to suppose him in danger. I am told that he was habitually imprudent in exposing himself to the most inclement weather, from an unfortunate theory of the benefit of hardiness; and certainly he did not look like a man who could, at 77 or 78, brave such exposure with impunity. Lord Cockburn tells me that the last thing Lord Jeffrey wrote (for publication) was the note at page 64 of the last number of the *Edinburgh Review*. The funeral takes place to-day; there was a strong wish that it should be public, but his family wished that none but personal friends should be present. One thing more I must add which I have heard—that he is almost as much regretted in his legal and judicial, as in his social capacity.

Fanny and I are reading Miss Martineau's "Egypt." It is extremely well written and gives one a very lively picture of what is to be seen in that strange country, but she is often too transcendental for me, and I cannot follow her enthusiastic rhapsodies about Osiris. Indeed I do not well understand her religious notions. She seems to me to rate too highly the civilization of the ancient Egyptians; she is too much captivated, I think by the mere magnitude of their works. And when she says that all our civilization and knowledge came from Egypt, she seems to leave out of sight the fact that the Egyptians were subject to as rigid a despotism, sacerdotal and royal as any other

eastern nation, and had as little notion of mental 1850.
liberty. A civilization derived wholly from them
would probably have remained as stationary as that
of the Hindoos. Political freedom and freedom of
thought were *invented* by the Greeks, and blessed
inventions they were. There is a passage quoted
from " Grote's History of Greece," at page 145 of
the last number of the Edinburgh Review, which
takes a very different view of Egyptian civilization,
and offers a striking contrast to Miss Martineau.

Perhaps Mr. Grote rather under-rates the Egyptians, yet I am inclined to think he is much nearer
the truth than the lady is.

I understand from a passage in your last letter to
Fanny, that you are continuing your researches
into the physical history of the Nile valley—a very
curious subject. Have you been able yet to arrive
at any conclusions? With respect to the coal
formation, I think the great question to be solved is
one, for a solution of which we must look to the
chemists—namely, what was the nature of the
process, and what the conditions by which ac-
cumulations of half decayed vegetable matter could
be converted into the hard, brittle, crystalline
substance we call coal. No doubt the *dynamical*
part of the question—relating to changes of level,
and the occasional alternations of fresh water and
marine deposits, offers plenty of difficulties ; but in
this part we must probably make up our minds to be
contented with plausible conjectures at the best ;
whereas the chemical question may admit of a
positive solution.

1850.   I have been very much pleased with Mr. Senior's article on Lamartine. You say that Lamartine was *dishonest*: this is not exactly the word I should use. Mischievous I think he was, for he was a visionary; he was a man who would sacrifice everything to a theory: and by his own showing, he was guided in his attempts at legislation by impulse and "instinct," not by sober reason. He did wrong, inexcusably wrong, in the first instance, in abetting the Revolution for the sake of a favourite theory; but it does not appear to me that he was ever influenced by personal interest or ambition: and therefore though I think his conduct culpable and mischievous, I should not call him dishonest. He showed great courage and patriotism in his opposition to the red republicans: and though I think him much to blame for not breaking off altogether from the Provisional Government when he found himself unable to control the Red section of it, I am willing to give him credit for an honest motive.

You see my letters are little else than dissertations, for in good truth I have scarcely anything to tell, unless I were to register the variations of the weather.

I dined on Saturday at Mr. Rutherford's—a pleasant party, although a gloom was at first cast over it by the news of Lord Jeffrey's death. Lord Cockburn is most kind in coming constantly to see us and chat with us; and I find him very agreeable when he is talking away thus at his ease in a quiet way. He has a remarkable fund of anecdotes and tells them peculiarly well.

Fanny is pretty well, though not making any 1850. very apparent progress.

I hope the question of the hours of labour will be effectually settled in this session. There are such important practical measures, especially sanitary ones, calling for the attention of parliament, that it will be a shame if the session is allowed to be consumed in mere party questions; yet I have my fears.

Pray give my love to all your party, — the united houses of Horners and Lyells and believe me,

Your very affectionate Son-in-law,

C. J. F. BUNBURY.

---

3. Rutland Street,

February 5th, 1850.

My Dear Mr. and Mrs. Horner,

A thousand thanks to both of you for your kind remembrance of my birthday, and your present of Humboldt's delightful book. I had indeed bought the book some time ago (a different translation however) and had studied it pretty assiduously but I am not the less obliged to you for so welcome a present; for it is a particular favourite with me. And I hope I hardly need say that every token of your regard and kindness is most truly welcome to me. The pretty book you have sent me will be an ornament to my shelves, and will serve for my own and Fanny's reading, while the copy I had before (which has a more Quakerish

1850. exterior) will do excellently well for lending.   My
darling Wife has made me a present of a beautiful
book, " Stewart's Active and Moral Powers."

I am very much obliged to Susan and Joanna for
their presents, of which the one is funny and the
other useful.   I will write to Mary separately.   I
perceive that she and Susan think I must be much
in want of something to divert me.   It is quite true
that events, whether comic or serious, are not over
plentiful here.   Still we have many pleasures and
resources, and I shall always think of this winter at
Edinburgh as a very happy time.

Yesterday was fortunately fine, and we enjoyed
together the beautiful drive round Arthur's Seat.
To-day we have a perfect storm of wind and rain.

Pray tell Leonora that I was at the Botanic
Garden on Saturday, and saw there a fine plant of
the *Stiftia chrysantha* in full flower.   I think she
has seen it at Kew; it is a great rarity in this
country (first introduced by Gardner), but is an
old friend of mine, as I gathered it in one of my
first walks at Rio.   There is also the Luculia
gratissima in full blossom, very beautiful and more
deliciously sweet than almost any flower I ever
smelt.

My Fanny is looking well.   She is very cheerful,
always well occupied, and makes me love her more
and more.   I cannot say what gratitude I feel to
you both for having given me such a Wife.

Much love to our Sisters.

Ever your affectionate Son-in-law,

C. J. F. BUNBURY.

Abergwynant,
February 7th, 1850.

My Dear Charles,

We have not heard *from* Henry by this Indian Mail, but we have heard of him, considerably to my relief: for I did not half like that fishing excursion, on account of jungle fever. M'Murdo very good-naturedly wrote from Lahore to Emily, to let us know that he had received a letter from Henry, who had arrived safe at Karigra, sound in all respects except as to his shoe-leather, which had perished in the snows of the Himalayah. He had succeeded in reaching a pass 12,000 feet above the sea.

I hope for very interesting letters by the next mail. Karigra is a fortress in the Jullunder Doab. It was there that Sir Charles Napier's escort and his train of followers were to assemble for his march to Peshawur.

Edward is in London, and makes a good report of himself.

I am, my Dear Charles,
Ever most affectionately yours,
H. E. B.

———

To Lady Bunbury.

3, Rutland Street,
February 12th, 1850.

My Dear Emily,

I thank you much for your letter of the 4th, which was, like most of your letters, very

1850. agreeable : and I was glad to find you writing in better spirits than my father, who seems to be sadly despondent about political matters. I dislike the democratic *tendencies* of the present times as much as possible, and do not wish to shut my eyes at all to the dangers which threaten our best institutions; but I have still too good an opinion of the English people to despair of the result. I am glad to find we are not to have a new Reform bill frenzy. We have bought a *lot* of Corn-law pamphlets, pro and con, but I have not read any of them yet, and do not know when I shall. Do you see in the newspapers that the Americans are speculating on buying up the deserted estates in Jamaica and other West Indian Islands, for the sake of growing *cotton* on them ? Creditable, certainly, to the judgment and enterprise of our West Indian planters. If we are entirely dependent on the U. S. for our supply of cotton, it is not, I suspect, the fault of nature nor of an inevitable destiny, but the West Indians (though I think they have been hardly used by our government) seem to have always had so much disposition to routine and to taking things easy, that one can hardly believe they came originally from the same race with the Yankees.

I know little about the self-constituted council for the Colonies. I only saw in the *Times*, several weeks ago, the sort of manifesto with which it began, announcing its formation and its objects, and an article in the *Times* (or rather the evening Mail) of the next day, criticizing the said manifesto pretty sharply : but I have seen no further notice of it.

I was struck, not very agreeably, with the strange 1850.
alliance of men of the most opposite parties in
the proposed Council. If I were in the place of my
old friend Stafford O'Brien, I would not coalesce
with the Cobdenites for any political object whatever.

You will probably have heard before this that
Mr. Rutherfurd declined the judge-ship vacant by
poor Lord Jeffrey's death, and that it has been
given to Mr. Maitland who was Solicitor General
for Scotland. I saw Mrs. Rutherfurd the other
day; she was cheerful and agreeable but not quite
well yet.

On Saturday I went with Professor and Mrs.
Forbes to see Hugh Miller and his collection of
fossils. Did you ever hear of Hugh Miller? He
is a very remarkable man: originally a working
stone-mason, with no other education than that of
the Scotch peasantry in general; he has made
himself a capital geologist (of the geology of
Scotland he knows perhaps more than any man
living),—his knowledge of fossil fish is consum-
mate; and his books, though provincialisms and
inaccuracies of the language may be found in them,
are written in a style singularly expressive, lively,
clear, and unaffectedly eloquent. One power he
possesses in a higher degree, I think, than any other
writer of the present day: that of illustrating his
meaning, when he treats of matters somewhat new
or difficult, by a singularly happy use of comparisons.
Unfortunately he is not merely a geologist and a
fossilist, he is one of the Free Kirk, and the

1850. conductor of its principal organ, "The Witness," in which I understand he pours forth a superabundance of theological venom. His collection of fossils is exceedingly rich and curious, and he did the honours of it very well. Mr. Forbes had told me that I should find him "as odd a fish as any in his collection," and indeed his appearance is rather Orson-ish, with his face almost hidden in a huge wild bush of red hair and whiskers, but there is nothing uncouth in his manner and conversation.

I quite agree with you as to the unsatisfactory moral tone of Gibbon's History : he seems, (in that work at least) incapable of feeling either a hearty admiration of the higher virtues, or a hearty abhorrence of wickedness, and this, and his fatiguing style, appear to me the only fault one can find with that magnificent work, which must otherwise rank among the very first of histories. In his private letters he appears in a much more amiable light. Fanny and I are both reading Sismondi's "Littérateur du Midi," which I read some 17 years ago at your recommendation, and which I find as I then did, a delightful book, although I dissent from some of his criticisms, and especially, as to the relative merits of Ariosto and Tasso. We borrowed it from our friend Mr. Erskine, who has been most kind in lending us books. He is really a charming old man : and Mrs. Erskine* too is very agreeable. I gave them your message, and they said there was no danger of them forgetting you.

Fanny is looking well and in good spirits, and I

* A daughter of Sir Charles Mackintosh.

hope is really improving in health, I trust she will 1850.
be well able to bear the journey to London. Much
love from both of us to dear Cecilia.

Believe me ever,

Your very affectionate step-Son,

C. J. F. BUNBURY.

P.S.—I have just been reading Lord John
Russell's great speech on the Colonies, and like
it much.

———

3, Rutland Street, Edinburgh,
5th March, 1850.

My Dear Lyell,

Many thanks for sending me your "President's Message,"*which I have read with care, as
indeed it well deserves ; it might be bound up with
your "Principles," to which it would form a valuable
supplement.

The facts brought forward in it are striking and
important, the argumentative or controversial part
very ingenious, and, like all you write, well worthy
of study ; though I scarcely think it will be considered as settling the question. It would seem
that the common phrase—"as old as the hills," is
(like many other proverbial phrases), founded on an
error, for Murchison and Darwin seem to show that
the *hills*, at least the biggest of them, such as the
Alps and Andes, are mere modern upstarts—tertiary
fellows,—creatures of yesterday ! "As old as the
plains," would be more to the purpose, since the

* President of the Geological Society of London.

1850. most ancient and venerable rocks make their appearance chiefly in the undisturbed levels of Russia
and the northern parts of America. To be serious,
I must own that I remain unconvinced by the
reasoning of the latter part of your address, ingenious as it is. But I must explain how far I am
a heretic. I have no more belief in a *chaotic state*
than you have; nor in *universal* convulsions or
catastrophes. As you say the undisturbed state
of the older rocks throughout extensive regions,
shows such a doctrine as that of universal convulsions to be untenable. But I cannot understand or
imagine how such gigantic effects as are visible in
the Alps and other great mountain chains, could
ever be produced by a succession of such trifling
movements as has been witnessed in historical
times. Nor do I see that the doctrine of "existing
causes" requires us to suppose that the igneous
agencies are never capable of greater exertions than
have been observed within the short period of
human record.

Only consider how very small a particle of geological time is that to which records of any kind
extend, whether historical or traditionary; and that
the period for which we have precise information
concerning the most important volcanic districts is
but a small fraction even of that particle. I doubt
whether we are quite justified in assuming that
no greater effects of volcanic power can ever have
occurred than such as are known to us from modern
observations. At the same time, I know that you
have Darwin on your side, and he is a host in

himself ; I can only say that I cannot conceive how 1850.
any series of such insignificant movements as those
observed in modern times (in Chili for instance)
could ever produce the upheavings, dislocations,
crumplings, foldings, inversions, etc., now seen in
the Alps and Andes. It is a mystery to me ; I
cannot *realize* it, as the Americans say. But to
discuss the subject fully would require a correspondence as voluminous as that of Cicero and
Atticus.

In the meantime I think your address is a valuable addition to the controversial literature of
geology, and look forward to the Second Part which
you promise, on the organic world, with which I
have no doubt I shall be able to agree much more
thoroughly. I have lately read over again the corresponding part (the 3rd book) of your "Principles,"
with very great satisfaction ; you have rendered
a most valuable service, not only to geology, but
to the geographical part of natural history.

With much love to Mary,

Believe me ever,

Yours affectionately,

C. J. F. BUNBURY.

3, Rutland Street,
March 16th, 1850.

My Dear Emily,

I dined yesterday at Mrs. Rutherfurd's ; a
very pleasant little party, the only guests besides

1850. myself, Mr. and Mrs. Primrose, and Sir John Acton.*
Mr. Primrose, a Son of Lord Roseberry, is a gentlemanlike, conversible man, and his wife is pretty and
agreeable.   Mrs. Rutherfurd seemed well, and in
good spirits, and the Lord Advocate was in
great force, and told many good stories.   He saw
you when he was in town.   I have been beside, at
two evening parties this week, but did not hear
anything very remarkable.

I have been much amused by looking over the
illustrated edition of " Jerome Paturot," which one
of our friends here has lent us; do you know the
book ? it is a clever satire on the late French
Revolution, and the illustrations are full of humour,
fancy and variety.

By the bye, matters seem to be again taking a
very bad turn in France; three Socialists have
come in for the Department of the Seine, and Mr.
Rutherfurd thinks that that worst of parties is
gaining a formidable strength.   Another French
Revolution, if it should end in the triumph of the
Socialists would throw all Europe into worse tumult
and confusion than ever, and would leave little but
the pleasant alternative of the predominance of
anarchy or of Russian despotism.   Whether this
country would pass unhurt through such a shock, I
cannot tell, but I feel no doubt that it will have a
better chance of safety without the corn-laws than
with them.

Till within the last three days, the weather has

* Now 1890—Lord Acton.

been very pleasant ever since the latter part of last 1850.
month ; several days really beautiful, showing the
scenery of this neighbourhood to the best advantage.
Certainly, Edinburgh has many merits and advantages.

With much love to Cecilia,

Believe me,

Ever your very affectionate Step-son,

C. J. F. BUNBURY.

17, Melville Street, Edinburgh,
March 22nd, 1850.

My Dear Emily,

I thank you heartily for your two kind
letters, and I feel very much your sympathy in my
anxiety and distress about my dear Wife, but she is
certainly much better than when I wrote to you.

We have fine, mild weather here, the east wind
not troublesome, and the season appears to me a
peculiarly early one ; the bushes are quite green,
the Ribes sanguineum and Berberis aquifolia very
nearly in blossom, and a variety of pretty flowers in
the open ground in the Botanic Garden ; particularly the lovely little Erythronium in full beauty.
Books and the Botanic Garden are my great
resources here. I am reading with great pleasure
Dugald Stewart " On the Active and Moral Powers
of Man," which Fanny introduced to my notice.
Do you know the book ? It is delightful both in
style and sentiments. As a companion to it I have

1850. also begun to read "Cicero de Officiis," to which (as far as I have yet gone) the same character may be applied with equal justice.

Much love to Cecilia. Oh, by the bye, do tell me if you have heard lately from Sir George; I hope he and Lady Napier are well, and as joyous as ever.

Believe me,
Your very affectionate Step-son,
C. J. F. BUNBURY.

17, Melville Street,
April 10th, 1850.

My Dear Mr. Horner,

I must write you a few lines to announce that there now really seems every probability of our setting off to-morrow.

I shall be *very* glad when we are safely arrived in London, as I cannot but feel rather anxious about the journey for Fanny. Yet as she did not suffer at all from our excursion to Bonaly the other day, I trust that with great care she may get well through it; and in that case the change of air and scene will probably do her good.

I have lately got acquainted (through the good offices of Miss Miller) with a man who pleases me much, so that l rather wish I had known him sooner,—Professor Smyth, the astronomer, who is son of Captain Smyth, the hydrographer. He was eleven years at the Cape of Good Hope (as assistant to Mr. Maclear), which is a great bond

of fellowship between us; for I am so fond of 1850.
the Cape that my heart warms to any one who has
been there. He has shewn me a great number
of interesting drawings which he made during his
stay; among others a very striking series of
sketches of the great comet of '43; and has given
me much information concerning the mountains,
climate, rocks, plants, &c.; for as he was employed
for a long time in surveying, and was encamped for
weeks together on the mountains, he had great
opportunities, and he seems to be a very good
observer.

We had a very agreeable visit to Bonaly on
Sunday, and were most kindly received, Fanny, as
you may suppose, was quite unhappy at parting
from Lord Cockburn, and I felt sincere regret, for
during our stay at Edinburgh, I have learnt really
to know and esteem and value, as well as like
him; and I must always feel grateful to him for
coming so constantly to see Fanny, and relieve the
monotony and weariness of her captivity.

I trust to find you in good health and spirits, and
to show you your daughter more blooming, at
least, than when you were parted from her.

Much love to all the House of Horner.

Believe me ever,

Your affectionate Son-in-law,

C. J. F. BUNBURY.

1850.   [I can find no written materials between April
and September. Our time was divided between
London, Mildenhall and Barton ; most of the time
was, I believe, spent at Mildenhall. In August we
went to spend some weeks at Scarborough, York
and Whitby].

F. J. BUNBURY.

Scarborough,
September 3rd, 1850.

My Dear Lyell,

I congratulate you on your safe arrival at
home after your flying expedition through Germany,
which seems to have been very successful. I have
read Mary's letter and your own with great interest,
though rather bewildered by the rapidity of your
proceedings, and the quick succession of scenes and
persons and ideas. Certainly you are the two
most active of human beings. It made me quite
dizzy to read of the quantity of things you saw and
did in each day, and I am sure that if I had
attempted anything like it, my brain would have
been in a perfect whirl without one clear idea left in
it.

We have been doing things here in a much
quieter way, but our visit to Scarborough has been
a very successful one ; that is I hope and believe it
will have been successful in the great point of
Fanny's health ; and it has certainly answered most
uncommonly well for me, with reference to my
scientific studies. I have learnt a great deal from

the collections here, principally Dr. Murray's and 1849. Mr. Bean's. Dr. Murray's kindness and liberality has given me the opportunity of drawing and carefully describing some new and unpublished things of great interest.

As it will be a long time before my book can come out,* I think I shall put the most important results of my study of these collections into the form of a paper for the Geological Society against this next winter. It is true the paper would be purely botanical, without any *direct* geological bearing ; but several purely zoological papers have appeared in our *Journal*. One thing which I have learnt, quite new to me, and not mentioned in any book that I have seen, is the existence of *true Calamites*, and good big ones too, in the sandstone of the oolite series of this coast—a sandstone which lies *above* the inferior oolite. They are evidently a different species from your Richmond calamites, as well as from the coal measure kinds, and are marked by the thick and tumid articulations. They are quite different from *Equisetum Columnare*.

Dr. Murray is exceedingly anxious to make your acquaintance if ever you should come into this part of the world, and I think you would be much pleased with him as well as with his collections ; he is a highly cultivated man, most courteous and liberal and communicative, seems thoroughly conversant with the geology of the neighbourhood, and evidently delights in conversing on scientific subjects. Mr. Bean, too, is an interesting person,

* This book, I am sorry to say. never came out.

1850. and his collections are wonderfully rich, both in fossils and in recent shells and zoophytes. Altogether, I think you would find Scarborough well worth a visit, if ever you were in the North again.

(Whitby, September 5th). We came here yesterday by railway, but are not much charmed (at present) with the place, except the new hotel (the Royal) which is excellent, and capitally situated. We have not yet seen the museum, but have visited the ruins of the Abbey, which are grand, but wretchedly neglected. I am much puzzled at finding the cliffs here, all consisting of sandstone, and wonder what is become of the lias !

We hope to be home on Tuesday, the 10th.

Much love to Mary.

Ever yours affectionately,
C. J. F. BUNBURY.

Mildenhall,
September 19th, 1850.

My Dear Lyell,

Many thanks for your very entertaining letter of the 14th. I am heartily glad that your tour answered so well in point of health as well as geology ; but for my part I should not have much enjoyed travelling through a country where the cholera was raging. Barring this, however, I should like very much to make a tour of the German museums, and we hope to execute such a plan, perhaps next summer, but at a different rate of progression from you ; for I belong to the *tardigrade* order of Mammalia. Besides, I should not like to 1850. make such a tour without leisure to look at the galleries of Art, as well as the museums and herbariums.

What I most envy you in your tour was that interview with Humboldt, whom I should above all things like to know. I hope he may be still alive when I go to Germany. Had you time when at Berlin to see the Botanic Garden? which is considered one of the finest in the world. It would be one of my chief attractions there.

We are both looking forward with great delight to seeing you and dear Mary here, and I shall have much to show you. Our garden is now in great beauty.

I am very busy and happy working at my collections, putting into shape and order my Scarborough notes, and cataloguing my fossils; while Fanny, besides her school and garden, has several literary enterprizes in hand.

Much love to Mary.

Ever yours affectionately,
C. J. F. BUNBURY.

Mildenhall,
October 8th, 1850.

My Dear Mr. Horner,

Since I received your letter of the 29th (which found me at Barton), we have been exceedingly shocked by the news of the death of poor Henry Nicholson. Fanny, having known him so

3 E

1850. long, and so well, was, as you may suppose, much
affected; and though my acquaintance with him was
but short, I had been much pleased with what I
saw of his character during our visit to Waverley;
and I feel much for his family, to whom it must
be a cruel and most unexpected blow.—We were
very much grieved at hearing of poor Lord Cockburn's danger; but the last account seems decidedly
better, and I trust he will get over it, though at
his time of life, the recovery from so severe an
attack must, I fear, be slow.

We spent a day-and-a-half at Barton with my
Father, who seems to be in very good health.

I am much in the condition that you acknowledge
in your letter,—of having nothing to say. I spend
my time very happily, and in a certain way very
busily, but without incident or much variety. I am
dressing up my notes on the Scarborough fossil
plants, for the G. S.,—arranging and cataloguing
my collections, and reading a little German. We
have both been reading "Shepherd's Life of
Poggio Bracciolini," and have been very much
entertained by it. Do you know the book? It is
not a new one; the author was a friend of Roscoe.
It gives an account of a lively and interesting time,
the period of the revival of classical learning in
Europe. It is one of the pleasantest biographies
I have read.

Believe me ever,

Your affectionate Son-in-law,
C. J. F. Bunbury.

My Dear Lyell,

Many thanks to you and dear Mary for your agreeable letter, which gave me great pleasure. I am very glad you both enjoyed your visit here; I am sure it was a very great treat to me, and I trust it will be repeated this summer; you really *ought* to see us in our summer dress, and besides, I am half provoked to find how many things (in my museum) I forgot to show you when you were here.

As Gray says that the Italians spend half their time in thinking of the last Carnival, and the other half in looking forward to the next; so Fanny and I are just now living upon the memory of our Christmas family party, and looking forward to the hopes of another. It has done us both a world of good.

Now to your question. The great mass of marine vegetation consisting of Algæ, is certainly lower in organization (or less developed) than the land vegetation. There are *some* flowering plants in the sea, though not many: one in our seas, Zostera marina; and several in the Mediterranean and other seas of the warmer temperate zones: species of Zostera, Caulinia, &c. These are endogenous, and according to Lindley are somewhat allied to the Arum tribe, but they are among the lowest in organization of flowering plants. They have perfect stamens and pistils, and seeds of the ordinary structure; but they have hardly any

1851. vascular tissue. With respect to fresh-water plants, the question as to their *rank* is rather less easy to answer. I presume you do not mean to include all fresh-water plants,—not the water-lilies for instance, which are as highly developed as any plants whatever. But to take those you name— the Charas have been so differently placed by so many eminent botanists, that it is not easy to say what degree of dignity ought to be assigned to them. Some of the greatest authorities,—no less than Linnæus, Jussieu, Decandolle and Rob. Brown have placed them among flowering plants; but I must own that I am of the opinion of Endlicher and Lindley, who consider them closely allied to Algæ. Ad. Brongniart places them near Equisetaceæ. It is certain that they consist merely of cel- lular tissue, and their fructification appears much more like that of some of the red Algæ, than real stamens and pistils. The water- starworts again (Callitriche) are unquestionably phaenogamous; and so are the Duckweeds (Lemna); the former exogenous, the latter en- dogenous, but both of a very low grade. Then there are the Pondweeds, Potamogeton,—endogen s of a rather higher degree of developement. On the whole, I think you may safely say that the fresh- water vegetation (even excluding those plants which rise *above* the surface of the water) includes a larger proportion of flowering plants than does that of the sea.

Thanks for your information about the red chalk: I will look out for it in all the pits I can find

hereabouts, and hope you will by-and-bye honour 1851.
Mildenhall with a report on its geology.

For a proof of the mildness of this extraordinary
winter,—the day before yesterday I saw a *Bat* flying
about in the open air, brisk and active as in
summer.

The thrushes have been singing delightfully for
some days.

The fossil fir cone of the coal measures is figured
at plate 164 of Lindley and Hutton under the
name of Pinus anthracina. Unger (from whom I
take the reference to L. and H.) calls it *Elate*
anthracina, considering it as like Abies rather than
Pinus.

Ever yours affectionately,
C. J. F. BUNBURY.

Mildenhall.
January 24th, 1851.

My Dear Leonora,

   I have never heard of any book upon the
Ferns of Spain, or of the south of Europe; nay, I
cannot recollect any on European Ferns exclusively.
The south of Europe generally is poor in Ferns; I
do not know whether Spain is an exception, but I
should hardly think it, from the accounts we have
of the dryness of the climate. Portugal, indeed,
has two very beautiful and remarkable Ferns—
Woodwardia radicans and Davallia Canariensis;
and it is possible they may grow in the south of
Spain also. I dare say you know them.

1851.   The Date Palm, I believe, is much cultivated in
the south of Spain, and grows to a large size, and I
rather think ripens its fruit there, but it is not a
native of any part of Europe.   The only European
Palm is the little fan-leaved kind or Palmetto, the
Chamærops humilis, which is said to abound on the
rock of Gibraltar.

Pray do not suppose it can ever be troublesome
to me to answer your questions, or to give you any
help I can in your botanical studies.   Of the
Brazilian ones that I gave you, those from Rio de
Janeiro and Gongo Soco, those from *Minas Geraes*
and those that are marked " Brazil," merely were
of my own collecting; those marked " Porto Alegre,"
" Rio Grande " and " St. Catherine's," were
collected by my Uncle ; those from Buenos Ayres
partly by him and partly by me.

Fanny is, as you conjecture, completely immersed,
absorbed, and—if I could think of any stronger word
I would use it—in school matters, and I wish I
could say that she is not working too hard.   She
has been much pleased with Mr. Morell, who dined
and slept here last night and went away at half-past
three this afternoon, after examining the school.

He seems a very clever man, and profoundly
learned in metaphysics and German philosophy ;
much too learned for me, for he was continually
getting beyond my depth ; but his manners are
pleasing, and I like his opinions on subjects that I
understand.   I think I should have known by his
appearance that he was a student of German and
metaphysics.

We expect the Dean of Hereford on the 3rd.   1851.

I have been reading Sydney Smith's lectures, which are very entertaining.   I thought the first two or three a little flippant, but have liked them much better since.

Believe me,

Ever your affectionate Brother,

C. J. F. BUNBURY.

To Lady Lyell.

Mildenhall,

February 4th, 1851.

My Dear Mary,

I thank you very much for your kind and pleasant letter, and your congratulations.   After one has turned the corner of forty, birthdays are not exactly what they are to younger people ; but if many more birthdays find me with as many sources of happiness and as little to complain of, as I have now, I shall have great reason to be thankful.

We have enjoyed the Dean's* visit very much ; he came about one o'clock yesterday, and stayed hardly 24 hours, but he was extremely pleasant, and has given Fanny great encouragement about her schools.   I like his manner and conversation very much, he is so frank, cheerful, natural and animated and so full of clear, strong, manly sense.   I was much interested this morning by his mode of examining and instructing the boys ;   it is quite

* Dr. Dawes, Dean of Hereford.   He was one of three who had lately been presented with Deaneries on account of the way they had promoted education, and *I think* they were called the *Educational Deans.*   The other two were Elliot, Dean of Bristol, and Hamilton, Dean of Salisbury. (F. J. B.)

1851. evident he has a genius for teaching. He spoke
highly of Mr. Phillips and seemed to augur excellent
things of the boys' school.

Indeed I must say myself that I was quite
surprised at the progress and improvement of the
boys since the last time I was there.

You know by this time what the Queen has said,
but we shall not know till to-morrow. To me it
appears that the object really important, in reference
to the " Papal Aggression," has been already
attained by the very fact of the agitation ; inasmuch
as it has clearly shown that England is resolved not
to submit to the Pope, or the principles of Popery ;
but I suppose something more will be required to
satisfy the popular feeling.

I look forward with great pleasure to Rivermede*
and London and expect to have great satisfaction
in reading the " Manual," which I hope will have an
immense sale.

I think the Dean's visit has really acted like a
cordial on Fanny, for she seems much better
to-day.

Ever your affectionate Brother,

C. J. F. Bunbury.

From his Father.

Barton,
March 5th, 1851.

Well, my dear Charles, this has been a curious
chapter in political history.

An unintelligible beginning and a very pitiful

* Rivermede, Mr. Horner's house at Kingston on-Thames.—(F. J.B).

ending, if ending it be. But I cannot believe that 1851.
the merely piling up again the same stones which
tumbled down the other day without any violent
shock, can have rendered the pile durable. There
was little feeling on the part of the public in favor of
Ministers before this lamentable exposure of their
weakness ; and now there will be still less. They
must compromise the matter with the Peelites some
how or other, or we shall have another of these
" Ministerial crises " before long. The only
satisfactory thing that I can see in what has
occurred, is the *honest* conduct we have seen in the
leading men of the political parties. Lord Stanley's
frank declaration of his views is also an important
matter.

I have a letter from Henry written just as he was
arriving at Malta. He* reckons on being in London
on the 11th or 12th.

We expect Edward here this evening, to pay his
duty to his constituents.

I am afraid there will certainly be a contest for
Bury, whenever Parliament be dissolved. The
Protectionists are in a *savage* mood.

Emily and I mean to be in town in the middle of
next week, to meet her brothers† and Henry.

Tell Fanny, with my love, that I thank her very
much for sending me M. Whitmore's clever
pamphlet, which demands a separate critique.

Always most affectionately yours, H. E. B.

---

* Major Bunbury had just arrived from India. (F. J. B.)
† Sir Charles Napier and Henry Bunbury had just returned from India.
(F. J, B.)

To Lady Bunbury.

Mildenhall,
April 9th. 1851.

My Dear Emily,

We have most ungenial weather to welcome us home : it is colder than at Christmas. Mrs. Horner writes of lovely weather at Manchester, —so it would seem that the *weather office* is as partial to the manufacturing districts as the other departments of the government ! We find all well here, and our house and garden in nice order ; and after six weeks of gloomy London lodgings, it is a pleasure to return to our neat, clean, spacious, comfortable rooms, with our books and prints, and "everything handsome about us."

I went to the Drawing Room last Thursday with the Charles Lyells and Susan Horner, and the Herman Merivales ; but I lost them in the throng, and Lady Napier and Minnie kindly brought me home. I saw Aunt Caroline looking very well, and Pamela, Catty, and Norah, in great beauty. It was altogether a very gay and brilliant scene and a great crowd. The night before I had a very good view of the Duke of Wellington at the American Minister's. I had not seen him *near* since the year '36 or '37. He looked very brisk for a man of his age. We had met the American Minister* at the Lyell's, and liked him ; he has a straightforward, hearty frankness of manner and a plainness of speaking, very unlike one's notion of a diplomatist. Both in

Mr. Abbot Lawrence.

manner and appearance he is different from other 1851.
Americans that I have seen, and quite like an
Englishman.

We are in great hopes of meeting Sir Charles and
Lady Napier and Minnie at Naples in the winter,
for we have pretty well made up our minds to winter
there. I hope dear Cecilia is going on well, and not
giving you uneasiness. Pray give our loves to her,
and believe me ever

Yours very affectionately,

C. J. F. BUNBURY.

Mildenhall,
April 11th, 1851.

My Dear Lyell,

I have been reading over your Anniversary
Address with great pleasure and interest, and with
great admiration of the clearness and ability with
which it is written. At the same time, ingenious
and powerful as is your pleading, I cannot say that I
think the question by any means settled. The
balance of evidence still seems to me to be against
the presence of the more highly organized classes
of animals in the earliest geological periods. It is
true that the evidence is as yet very incomplete,
but such as it is, it seems to be all one way, and
you have nothing to oppose to it but conjectures :—

1.—You have very well pointed out the peculiar

1851. circumstances of the localities in which the coal
measures seem to have been formed, and given a
very proper caution against taking them as types of
the then condition of the whole earth.   Here it is
that I think the theorists are in error.   I think
the evidence we have is quite sufficient to bear
out Ad. Brongniart's views as to the nature of the
carboniferous vegetation in those *particular districts
where the coal was formed ;* I think we are justified in
concluding that the vegetation of *those districts* was
made up almost entirely of Acrogens and Gymnospermous Exogens ; but it does seem rash to
assume that there was then no different vegetation
anywhere on the face of the earth.   The climate
and other circumstances of the carboniferous
districts must have been *very* peculiar and extraordinary, for though we find at the present day a
certain approximation to their vegetational
characters in the islands of the southern hemispheres
yet I do not know of any island or tract of country
now existing, in which the two classes of plants I
have mentioned predominate to the *exclusion* of
Exogens proper.

2.—You express your dissent (p. 39.) from
Professor Heer's inference, that insects were very
rare in the carboniferous period owing to the
scarcity of flowers in those forests, in which Ferns,
Lycopodia and Exquiseta predominated.   I cannot
quite agree with you here : at least it seems to me
reasonable to suppose that vegetable-feeding insects
would have been comparatively scarce in the carboniferous forests.   The Conifers, indeed, would

doubtless afford nourishment to. many of the timber- 1851.
eating kinds, but no insects that we know of feed
upon Ferns or Lycopodiums. And this suggests a
curious question, which I have never seen touched
—namely by what agency was the superfluous and
decaying vegetation of that period removed ? In
the thick and damp forests of warm climates, at
the present day, fallen trees and dead vegetable
matter are rapidly cleared away by insects,
but this it would seem, would not have been the
case where the mass of the vegetation consisted of
Ferns and Fern-like plants. At any rate those
numerous tribes of insects which feed upon the
juices of flowers, must probably have been
wanting in the flowerless forests from which our
beds of coal originated.

3.—In speaking of the supposed Endogens of the
coal formation, I was at first rather surprised that
you had not noticed the Zeugophyllites which is an
undisputed Palm, from the coal field of Bengal ; but
I find that Brongniart does not enumerate it
among the plants of the coal period, considering
the age of the deposit to be doubtful. If that
Indian coal-field be really Palæozoic, the occurrence
of a true Palm *there* seems to dispose of the
developement theory as effectually as if it occurred
in Europe. When the freshwater and estuary
formations of tropical countries come to be well
examined, we shall probably know much more
of the old world history of Palms than we now
do.

4.—Although a few instances are now known

1851. of the occurrence of reptiles, or their tracks, in the coal measures, still I think appearances indicate that they were scarce; nor would this be surprising if the climate was such as I conjecture it to have been; for reptiles seem to be scarce in those islands of the southern hemisphere where the conditions approach nearest to what I conceive to have existed in the carboniferous lands. Excessive moisture in a *hot* climate may be favourable to reptile life, but in a temperate one it seems to be rather the reverse.

5.—At p. 58, as I understand you, you adopt Forbes's hypothesis, that the introduction of the existing species of animals and plants from Europe into this country, must have taken place before the separation of Britain from the Continent. This is questionable; most botanists, I observe, are of opinion that his assumptions on this head are unsound, and his bold hypothesis not required to explain the facts. Starting from the same hypothesis of specific centres, they hold that so narrow a sea as that which separates Britain from the Continent would not be a barrier to the immigration of most plants. Whether the same may be said of animals I do not know,—probably it may be more doubtful.

These are all the remarks that occur to me at present, and I hope you will not think them unreasonably long.

We were very glad to hear of your having a pleasant dinner party at the Palace.

I suppose you will send an abstract of your

lecture to *The Athenæum;* if so, pray get me an 1851. *Athenæum,* and send it down to us.

With much love to Mary, I am ever,
Yours affectionately,
C. J. F. BUNBURY.

To Mrs. Henry Lyell.

Mildenhall,
May 8th, 1851.

My Dear Katharine,

I thank. you for your pleasant letter of the 4th, and I was very glad to hear such a nice account of you and your dear little boy. I have just finished the notes on Cape Botany, which I have been writing for your use, as a kind of "botanist's guide;" though in truth it is hardly necessary, for in the season when you will be there, a botanist can hardly go wrong in that country; in whatever direction you go out of Cape Town you are sure to meet with a multitude of attractive and interesting plants. But it has been a pleasure to me preparing these notes, both for your sake and because I delight in every occupation that revives my recollections of that place and recalls the delightful days I spent there. I hope you will be able to stay there long enough to make several excursions. I expect to get a great deal of interesting botanical information from you about India, especially if you visit the hills.

The two Bauers were really wonderful men in

1851. their line; though there have been and are many excellent botanical draughtsmen, none have ever approached *them*.

We were delighted with Susan's capital account of the opening of the Exhibition, which must indeed have been a wonderful and interesting sight; I am very glad you were able to see it so well. How prosperously it all went off, without anything that could the least mar the pleasure of it. This is really something of which we may be proud, on behalf of our people and government, as well as thankful.

We are going to have a little Exhibition (a very little one) here; for the "Bury and West Suffolk Archæological Institute" (there is a grand title!) is going to hold the next meeting at Mildenhall early in June, and we give them the use of the school hall for the meeting and for the exhibition of antiquarian curiosities.

I hope to see you all again in little more than a month, and I trust we shall then be able to make out our expedition to Kew, and see some Cape plants there, though by the way, the gardens are poorer in them than in the plants of almost any other country.

Have you seen Joseph Hooker? You will doubtless get from him an immense deal of information about India. If you see him, pray remember me particularly to him.

Ever your affectionate Brother,

C. J. F. BUNBURY.

Mildenhall,
May 27th, 1851.

My Dear Susan,

I do thank you most heartily for the gift of your charming picture* which has arrived quite safe, and has perfectly enraptured us both. It is really beautiful, I hardly know any picture you have ever done that I like better: both the composition and colouring are to my taste, excellent, and it has the further merit of strongly reminding us of bright sunny Nice.

Ever your affectionate Brother,
C. J. F. BUNBURY.

To Lady Bunbury.

Mildenhall,
June 3rd 1851.

My Dear Emily,

I dare say the woods of Abergwynant are charming with the spring flowers, and the luxuriant growth of moss forms a beautiful *setting* for the smaller plants. I am very fond both of the Adoxa and the Wood-sorrel, the former *used* to grow at Barton, but it is very long since I have seen it alive ; the Wood-sorrel I never saw in Suffolk. Are your Rhododendrons very fine this year? We have found our garden in excellent order and condition and doing great credit to Elemer, but till to-day, we have had no sunshine to show it off.

* A Girl and her Donkey, taken at Nice in 1847, and the picture made from the sketch.

1851.   My old friend Harvey, whom you have often
heard me talk of, is going out on an expedition to
Australia and New Zealand, to botanize, and
especially to collect sea-weeds, of which he has
made a very particular study.   Do you know his
pretty little " Sea-side Book ? "   By the way, there
is now an interesting sight to be seen in the
Zoological Gardens, where they have got salt-water
tanks, in which a variety of those strange creatures
that live at the bottom of the sea, and which it is so
difficult to get a sight of, are living and thriving,
and may be seen at one's ease through the glass
sides of the tanks.   There one may see star-fish,
sea-urchins, sea-slugs, sea-anemones of every colour,
all sorts of strange crabs, barnacles and other shell-
fish, all alive, walking, crawling, wriggling, jumping,
swimming, spreading out their feelers, as if quite
athome.   It is really a sight worth seeing, and
few things that I saw in London amused me
more.

Some time ago I sent Harvey a large parcel of
dried plants selected from my Uncle's collections,
and he has sent me in return a magnificent set of
exotic sea-weeds (many of them from the Antipodes)
and of North-American plants, which will furnish
me with plenty of occupation in arranging and
studying them.

I have got a capital new microscope* made for me
under Dr. Hooker's directions, with which I *intend*
to do great things ; but unfortunately my doings are
apt to fall rather far short of my intentions.

* He gave it to his Nephew, H. C. J. Bunbury.

What do you think of Sir Charles Wood's India 1851. plan ? It strikes me at first sight as a sort of make-shift, which will satisfy nobody. I have my doubts whether it will be carried, but I am not in the way of hearing anything.

Our friend the Dean of Hereford is coming to us to-morrow, and the next will be an important day, as he is going to give our schools a regular examination.

I hope you have now pleasant weather in Wales, and are well enough to enjoy it. To-day (the 5th), it is really warm here, and the wind has got round to the west. I hope Cissy continues well. Pray give my love to her as well as to my Father and Henry, and believe me ever

Your affectionate Step-son,

C. J. F. BUNBURY.

[NOTE.—I can find no materials between these two dates. I believe our time was spent quietly at Mildenhall, except occasional visits to Barton].

F. J. B.

To Mrs. Henry Lyell.

Mildenhall,
November 2nd, 1851.

My Dear Katharine,

I was very happy indeed to hear of your safe arrival at the Cape, and very much pleased and interested by your letter, which I took as a graet

1851. mark of kindness, considering how much you must have had to do.

I am very glad you were so much pleased with the flowers at the Cape, as indeed I was sure you would be, arriving at such a favourable season. A visit to the Cape is indeed a glorious treat for a botanist, and one never to be forgotten ; and then the climate is so favourable to exercise, and one can explore with such perfect safety, and so free from the annoyance of noxious insects, and other usual plagues of hot countries, that I have never seen anywhere, such a Paradise for botanists. I am afraid you will find a very different state of things in India. I shall be very curious to see your Cape collection, and, though made in so short a time, it will very probably contain many things which escaped me ; for in one year I cannot have collected one-half, nor probably one-third, of the plants even of the immediate neighbourhood of Cape Town.— I agree with your remarks, too, upon the Cape in other points of view ; to a person visiting it without the advantages that I had, and without a botanical taste, it would probably appear a very dull place. It is quite true that man has done comparatively little for it, but when I visited the Cape, I had Brazil fresh in my mind, and naturally compared it with that, the only other extra European country I had seen ; and the same observation is so much *more* true with reference to Brazil, that the Cape appeared by comparison, an improved and improving country. I shall always love the Cape ; the year I spent there was perhaps the happiest of my

bachelor life, and everything that recalls the images 1851.
of it is pleasant to me. To revisit the place now,
would be a pleasure alloyed with much pain; yet
when I read your letters, I could hardly help
wishing that Fanny and I had been with you, and
that I could have acted as your guide, and ascended
Table Mountain once again. But do not suppose
I am discontented, for it is impossible to be happier
than I am; my life is really one of constant sunshine; but the "vestiges of the old flame"
(botanical, I mean) will sometimes blaze up again,
and make me a little impatient of our scanty
Flora.

Of ourselves, I fancy I can tell you little but what
will be in Fanny's letter, for I know she is
writing to you. She is very busy preparing for the
press her translation of Balbo's Life of Dante,
which Bentley has undertaken to publish without
risk to her; two proof sheets have already arrived,
and we worked together yesterday and the day
before, at correcting them. I am very much pleased
that it is to come out, and I think it will do her
great credit. The correction of the press, however,
delays our departure, and I suspect it will be more
than a month yet before we shall leave England.
We are going through France, and shall stay some
days at Paris; then embark at Marseilles for Malta,
which will be our head-quarters, but I hope to
get as far as Athens, and to stand on the Areopagus.
How we shall come home is as yet unsettled,
and indeed must depend in some measure on the
state of Europe, for it is quite possible that there

1851. may be a. grand *blow up* next spring, and though
Fanny might perhaps like to take part in a
Revolution, I do not wish that she should run
the risk.   She is looking very well and growing
fatter, and is certainly stronger than last year, but
(the old but) her mind is too active and eager for
her bodily strength.

We were both delighted with your accounts of
dear little Leonard.   For myself, I have been
writing a short paper for the Geological Society :
am working busily at South-American plants ; and
am going on leisurely with my book on fossil
plants, which may perhaps be nearly finished by the
time you return from India.

I suppose you will by this time be near Calcutta,
and before you receive this letter you will be far up
the country.   I understand that Harry's regiment
is ordered to the Punjaub, so on your way thither
you will see a great deal of India, and doubtless will
see much that is remarkable and interesting.   I
hope you will not suffer from the fatigues of
the journey, and above all I trust that there will be
no war.

As you so kindly offered to collect plants for me,
I may as well mention what I should like particularly
to have, provided you have the opportunity of
procuring them.   First :—Ferns and Mosses ; but
unless you are able to go to the hills, you will meet
with few of these.   Secondly :—Specimens (twigs
and leaves) of the Banyan and the Peepal tree.
Thirdly :—Any Leguminous plants you may meet
with, whether trees, shrubs or herbs ; noting par-

ticularly that in that family of plants, the fruit is of 1851.
much more botanical importance than the flower.*
Fourthly:—Specimens of any plant extensively
cultivated in India and not in this country; in particular, I should be very glad to have specimens of *Rice*
and the *Indigo* plant in flower and fruit. Fifthly:—
Separate fruits of any kinds of Palm you may chance
to meet with (except the common Cocoa-nut), and
any other fruits of remarkable form or character,
and of a nature to be easily preserved dry. This
list is intended merely as a sort of guide for you,
not at all with a wish to give you any trouble; and
you may be assured " that the smallest donation
will be thankfully received." Only let me urge you
not to fall into the error of inexperienced collectors,
in making your specimens too small and fragmentary. In the case of herbaceous plants (not
climbing) it is important to have the lower part
of the plant with the root leaves; and with respect
to shrubs and plants too large to be gathered
entire, I have found it a good rule to gather
specimens *as large as my drying paper would hold.*

There are however, some hard, scrubby, thick-
branched things, with which it is impossible to
follow such a rule, and which indeed one cannot
make into good specimens any way. I hope you
will keep a copious journal, and I am sure your
observations will be very interesting.

You will be delighted to see in the newspapers,
what a reception Kossuth has had in England, in
spite of the impotent venom of *The Times* which has

* To be remembered.

1851. been striving with all its might to write him down. His language and demeanour in this country seem to have been very reasonable and judicious, which to say the truth I hardly expected after his address to the democrats of Marseilles.

May God bless and protect you, dear Katharine, and bring you safe back to us.

Ever affectionately yours,
C. J. F. BUNBURY.

4, The Grove,* Highgate,
Dec. 26th, 1851.

My Dear Father,

I thank you very much for your letter of the 18th, which I should have answered much sooner, but that I wished to be able to say something decided respecting our movements. In consequence of what you there said, I thought it right to re-consider the whole question very carefully, especially as Fanny herself had some fears about the voyage. I assure you that I would on no account think of undertaking such a thing, however interesting the tour might be to myself, if I thought there was any risk of its doing her harm; I have in fact left the decision very much to herself, and was perfectly willing to have given up the whole plan.

Fanny herself could not make up her mind to give up the scheme of going abroad, for which we have

* The house to which Mr. Horner had removed from Rivermede.

so long been preparing. We have therefore at last 1851.
resolved upon going, and we shall take our passage
by the steamer of the 7th. At present, the weather
is very fine and calm, and seems likely to turn to
frost, which would suit us very well; but if, as
the time approached, it should threaten to become
very stormy, I should not hesitate to sacrifice our
passage-money.

I am very glad that our opinions on the
usurpation in France agree so well, and I
read with great interest your speculations on the
probable course of events. The French are so
strange a people, so difficult to understand or to
reckon upon, that it would be rash to attempt to
foretell anything as likely or unlikely to happen
among them; but if they *do* submit quietly to a
tyranny so far surpassing anything they have experienced since 1814, it will certainly be one of the
strangest phenomena of our strange times. I do
not think the voting will prove much, because in the
first place, us you say, the votes are likely to be
falsified by the agents of the President, and
secondly, many of his enemies will doubtless think
it useless and inexpedient to oppose him in *that*
way, and will reserve themselves for another occasion.

Louis Napoleon seems to be making advances
to various parties in various ways, and especially
to be courting the *parti prêtre;* but if we may
believe the Paris correspondent of the "Express,"
the measures of finance that he meditates are such
as will stamp him at once as a Socialist and utterly

1850. alienate all the monied and *proprietary* classes.
I hear however that the said correspondent is no
other than Dr. Dionysius Lardner, a person not
particularly deserving of confidence.—Sir Charles
Napier and Sir William take up with great zeal the
cause of Louis Napoleon,—which I do not wonder
at, considering their predilections, but Henry
Napier takes quite the opposite side.

You will see in the newspapers that Lord Palmerston has resigned and Lord Granville is appointed
in his stead; but I have not been able to learn
anything beyond what is in the papers. Edward was
here yesterday evening, but could give us no information, either as to the causes of this change, or
as to what may be expected to follow. For my part
I am sorry Lord Palmerston is out. I looked upon
him as the most valuable member of the Ministry.
I fear that the despotic powers will look upon his
removal as a triumph for *them;* and I almost fear
that a less determined and more compliant tone
may be adopted in our foreign policy.

I cannot help fearing much for poor Piedmont.
France on the one side hankering after Savoy,
and Austria on the other longing to destroy whatever
remnants of freedom are left in Italy,—is there
not great danger of a coalition for the purpose of
plunder? And ought we in such a case to look
tamely on? In truth, I can sometimes hardly help
wishing for war, because I see little chance of the
preservation of peace without the utter extinction
of liberty on the Continent.

I am very glad to hear from Henry that

he has a prospect of obtaining an unattached majority, and that he does not propose to leave the Army.

Edward goes down to Hardwicke to-day. He has heard that a dissolution is expected to take place very soon.

We shall be very much obliged to Emily and to you for letters to Sir Richard Pakenham and Sir Robert Gardiner.

I wish I could see you again before we leave England, but I trust to find you all well and happy at Barton next summer.

Pray give our loves to Emily and Cecilia, and believe me ever.

Your very affectionate Son,

C. J. F. BUNBURY.

P. S.—Miss Hallam is going to be married to a Captain Cator,—a very satisfactory marriage as I hear.

[NOTE.—It was finally decided that we should not go abroad. We first gave up the journey through France on account of the *coup d'etat* of Louis Napoleon, which rendered travelling through that country inconvenient. We next decided to go by sea to Gibraltar and Malta, but he gave up the passage on account of my fears, although we had taken our berths in the "Madrid." It was a very stormy season, and we soon after our decision

1851. heard that that vessel encountered a terrible storm on the North Coast of Spain and was nearly lost. It was the same season that the Amazon was lost].

F. J. B.

Ventnor,
February 6th, 1852.

My dear Mary,

1852.   Many thanks for your agreeable letter and kind remembrances of me, and for the very pretty little purse, which Fanny says is quite large enough for as much money as I *ought* to have at once! You see how tight a hand she keeps on me. I did not receive these things till to-day, as the bad weather induced us to stay at Ryde two days longer than we had intended. It would really have been a pity to come hither in bad weather, for our drive to-day has been quite delightful, and the scenery, which without being magnificent or astonishing, is remarkably pretty, varied, and interesting, was shown to great advantage by the changeable day, the bright sunshine and shifting shadows of the clouds.

On Tuesday we made a charming excursion from Ryde, Fanny in a Bath chair and I on foot; we visited Binstead quarries and Church, went as far as Wooton Bridge and returned by a circuitous route through a very pretty wooded and varied country. The small size of the Island perhaps enhances the beauty of the scenery, for one cannot go far any way without catching a peep of the sea, generally in

charming glimpses between the hills or down some
wooded valley.   Dr. Mantell's little book, which
I borrowed from Mr. Horner, is a capital guide
to the geology, and there is so much variety in the
formations within a small space, and such excellent
sections, that it would tempt almost any one to be a
geologist.

I have not done much yet in the way of collecting,
but at Ryde I found out Mr. Fowlstone, whom
Mantell recommends, and bought from him a fine
specimen of fossil wood and two pyritized cones
of Mantell's Abietites Dunkeri,—all this from the
Wealdon of Brook Bay,—and a specimen of freshwater limestone, containing seed-vessels of chara.

I was interested by the quarries of freshwater
eocene limestone at Binstead, where we bought
a few fossils from the quarrymen.

The section of the whole cretaceous series at
Sandown, which we passed to day, is very conspicuous and striking, even in a distant view.   I
hope before long to examine it in detail.   One
cannot expect to find anything new in the Isle
of Wight, which has been so well explored by so
many first-rate geologists, but I hope to increase my
practical skill and knowledge of the subject by
working here.—It is the best season too for the
purpose, for I can see that if I was here in the
summer there would be so much botanical attraction
as would leave me no leisure for geology.   I ought
to have written some time ago to thank Lyell for
sending me his supplementary sheet, which I received at St. Leonard's, and which I read imme-

1852. diately and with much interest; but in moving from place to place, I forgot to acknowledge it.—The additional facts he brings forward are very curious and very clearly stated, though I must own that I still hesitate to adopt his theories to their full extent; but I have not time to discuss the question just now.—I rejoice to hear that he is going to give a lecture on the 2nd April, and hope certainly to be present at it.

Much love to him from both of us.

Ever your affectionate Brother,
C. J. F. BUNBURY.

Ventnor,
February 6th, 1852.

My Dear Mrs. Horner,

I thank you very much for your kind letter and good wishes on my birthday, and for your very pretty and useful presents.

We had a delightful journey hither from Ryde to-day, travelling in a light, open carriage. We stopped first to see Brading church, which is old and quaint, with some curious monuments, and a sexton not perhaps *quite* as old as the church, but as quaint and primitive looking. Brading itself is a remarkably old-fashioned village and boasts of a *pair* of *stocks*, now a rare sight in England. Next we stopped at Yaverland to see another curious little old church, and again at Shanklin, were we went down to the shore, and enjoyed a fine view.

Ventnor is a very pretty place, in a strikingly 1852. picturesque situation; but is a much larger place than I had at all imagined—quite a town. We are at present in a very nice hotel (the Royal) delightfully situated.

We are both of us in love with the Isle of Wight, and I think if we have tolerably favourable weather during this month, it will make ample amends for our uninteresting fortnight at Brighton, during which my only intellectual resource was the reading-room. I believe, however to say the truth, that we both profited in bodily health by the air of that place,—though I have not the least wish ever to see it again.

Dear Katharine's letter from Allahabad was delightful, and I do hope and trust we shall continue to have comfortable accounts from her. I hope Susan is by this time safe in England again,* as she wrote that she was to cross on Friday if the weather should be favourable, and it is now a lovely moonlight night without a cloud. It is getting so late that—having being in the open air all day—I really must go to bed.

Fanny is looking very well, and in excellent spirits.

Much love to Mr. Horner,

Believe me ever,

Your affectionate Son-in-law,

C. J. F. BUNBURY.

---

* She had been staying at Paris with her Aunt, Mrs. Power, who had lost her Husband.

1852.   Ventnor,
                                    February 8th, 1852.

My Dear Leonora and Joanna,

I hope you will excuse my writing to you jointly, as I have hardly matter enough for two separate letters, and I do not like repeating. I thank you both for your kind notes and good wishes on my birthday, and for your pretty card-case, which is delightfully sweet, and will certainly keep my cards, if not my memory, in good odour. We are very much pleased with what we have seen of Ventnor, and expect it to be a very agreeable head-quarter station. Yesterday the weather being beautiful, we had a delightful ramble along the undercliff, for about 4 miles westward; the scenery is extremely pretty; the bold and continuous cliff which runs for many miles, like a great wall, at the back of the undercliff, and shelters it from the north, and from all cold winds is very picturesque and striking, and the sunny terrace of the under-cliff itself is beautifully varied with woods, pastures, rocks and gentlemen's seats. The woods which in some parts slope up to the foot of the cliffs are charming; the luxuriance of the vegetation, and especially of the Ferns and Ivy, is beyond anything I have seen elsewhere in England. The Harts-tongue, which is much the most abundant Fern here, grows in such masses, and of such size and beauty, that it might be taken for the Indian Asplenium Nidus; and the Clematis mantles the trees from top to bottom in a way which remind me of tropical climbers. The Iris foetidissima is one of the most characteristic plants of the Isle of Wight, 1852. abounding everywhere in the woods and hedges, where its large tufts of bright green leaves, and showy orange-coloured seeds are very conspicuous at this season. We have found also abundance of the wild Madder, Rubia peregrina, which is quite a southern plant. In the gardens the mildness of the climate is indicated by the luxuriant growth of the Evergreens, by the Coronilla and Stocks in full blossom in the open ground, and by the Passion Flower both in blossom and in fruit against the walls of houses.

The geology of the island is very interesting, and with the help of Dr. Mantell's book, which is an excellent guide, I hope to make myself well acquainted with it—though I can hardly expect to make any discoveries. I am as yet a very raw beginner in the art of fossil hunting, and should be the better for some of Joanna's lessons, as she gained so much experience in Belgium. We made two visits to the quarries of freshwater eocene limestone at Binstead near Ryde, got a tolerable knowledge of the characters of the formation in situ, and bought a few good fossils from the quarrymen. On our way hither we had a passing view of the section in Sandown Bay, which I hope to examine more in detail and at leisure another day ; the anteclinal dip of the stratification is very conspicuous, and the contrast between the white cliffs of chalk, the blackish gault and the dark feruginous brown of the lower greensand or neocomian formation strikes the eye at a distance.

1852.   Yesterday we examined the upper greensand, which forms the long line of cliff I have mentioned at the back of the undercliff, it has a very remarkable appearance, for the beds are of very different degrees of hardness, and are very unequally acted upon by the weather, so that the harder beds form bold prominent ledges, which extend continuously a great way along the face of the cliff, and afford famous nesting places for hawks and crows. We collected a few small fossils, but I was ill-provided with tools as well as with skill.

I hope dear Susan is before this, safe at Highgate; and now that she is on the safe side of the water, and can speak freely, I shall be curious to know what she really thought of the state of things at Paris. I have not yet read the *Edinburgh Review* article on Reform, but I infer from the allusions to it in the newspapers that it is very moderate ; otherwise I should have feared that your notions and mine on Parliamentary reform were hardly likely to agree.

Much love to Mr. and Mrs. Horner, and Susan, and my *best regards* to Thistle and Polly.

Fanny is very well and sends her love.

Ever your affectionate Brother,
C. J. F. BUNBURY.

Barton,
February 8th, 1852.

My Dear Charles,

I have to thank you for your letter of the 4th, and I am rejoiced to hear so good an account of Fanny's health. And now to say a few words in answer to your repeated inquiries as to my opinions regarding the Cape war.

The great Duke has just told us that Sir Harry Smith's measures have been admirable and unobjectionable, and has added the way to finish the war successfully, is to make roads through the strongholds of the Kaffirs. With all the deference due to such high authority, I venture to think that his judgment as to his general is erroneous, and his proposal impracticable. The plan of carrying on the war by "Patroles" that is by moveable columns traversing the Kaffir's country, is not Sir Harry's own. Maitland, and afterwards Pettinger prescribed it, and Smith has adopted it. It may be readily conceived that while the Kaffirs were occupying their Kraals, cultivating their millet, and pasturing their cattle over the country, they might be reduced to distress, and perhaps to submission by a system of continued disturbance and attack. But Sir Harry has presumed in acting upon this plan where there were no Kaffirs, or cattle or crops in the open country, and when his enemies had thrown themselves into the bush and kloofs on the very frontier of the colony (90 miles in rear of our General's headquarters) and were living in plenty on the sheep

3 G 2

1852. and cattle of the Colonists. This becomes intolerable, and then Smith sends inadequate bodies of troops to attack first the Bush, and afterwards the Water Kloof. He is foiled, not to say beaten, in both instances, and the only result is that the savages have made much progress in the art of war, and have acquired great confidence in themselves, and in their fortresses. My opinion is that as soon as that change of circumstances took place (I mean when the enemy abandoned " Caffraria," and threw themselves in masses into the Amatolas and the fortresses on our frontiers) Smith should have adapted his plans to the change, and abandoning Williamstown for a time, have gathered all his forces on our own borders, and applied his *whole* strength to the blockading, or if necessary to the scouring out of the enemy's strongholds. But what strikes me as the greatest of blunders has been this, Sir Harry had as much or more than he could manage upon his hands with the Kaffir tribes alone; yet at the same time, he or his deputy, provoked quarrel with the other Blacks in the sovereignty, who had shown no unfriendly disposition towards the English, and who may prove formidable enemies. Somewhat in the same light I view Sir Harry's present expedition across the Kei. Arabi and his people may be treacherous and hostile in their hearts; but as yet they had not used their arms, and it was our business to temporize and pretend to shut our eyes till we had finished our dealings with the Gaikve and the Pelambior. As it is we shall have renewed the pressure of war from our

most formidable foes for a time which they may 1852.
turn to account by ravaging Albany and Somerset;
we shall have knocked up our cavalry, and then we
shall brag of having taken so many head of cattle,
and shall return to our former position, with this
difference that the country from the Buffalos to
the Kei will be swarming with active warriors.

To turn back to the Duke's scheme of making
roads. How is it possible to make roads through
the Tirsko Bush? How can one deal with the
succulent plants of which it is composed? Fire
cannot be used, and mere cutting would be so slow a
process that the plants would be growing up again
before the workmen had half finished the clearance.
In such tremendous rifts of rock as the Water
Kloof and its neighbourhood, a road would just
bring one to a waterfall or a precipice; and what
would be the rate of insurance on the lives of the
engineers who should mark out, and of the workmen
who should makethe road? It would require half
the army to protect their work. Nor if the description I have heard of the Water Kloof and Blink-
water clusters of rifted rocks be correct, would
roads be effected in that locality? I could say
more, but I am at the end of my paper, so love to
Fanny, and God bless you,

Ever affectionately yours,
H. E. B.

                              Brook Street, London, Hathaway's Hotel,
                                          March 15th, 1852.

My Dear Mr. Horner,

I owe you many thanks for your kind letter, which I received nearly a fortnight ago—so fast does time fly ! I am happy to be able to give you a good account of my dear Fanny : she is looking remarkably well. We left the Isle of Wight with real regret, for we had passed a most agreeable month there, and few parts of England have pleased me more. The south coast of the Island is charming, and though the season was not perhaps the most favourable, I found very little interruption to my researches from weather, and was out of doors the greatest part of almost every day, finding constant occupation in marine botany and in geology. That coast is certainly the finest practical lesson in geology that I ever saw, and seems as if it had been framed by Nature expressly for the purpose of training young geologists. I have heard that Sir John Herschel said—"The Isle of Wight would make a geologist out of a stone;" and indeed I think it almost would.

I collected a good many fossils, and in particular some very fine specimens of fossil wood (coniferous), with the rings of growth, the course of the fibres, and the operations of boring animals beautifully exhibited. The cold winds that prevailed latterly prevented me from visiting some of the geologically interesting points, such as Compton Bay and Brook Point. We spent a delightful day at Alum Bay, which interested me exceedingly, but I could willingly have employed several days in studying 1852. its phenomena.

During the latter part of our stay at Ventnor, we had a very pleasant little society. Minnie Napier's company added much to the pleasure of our tour, and it was with regret that we parted from her at Winchester.

We are very comfortable at this hotel, and I am enjoying my stay in town exceedingly, thanks particularly to the Athenæum, with which I am quite delighted. I spend some hours of every day in its library, which is an immense resource to me, and I have already made great use of various books that I find there; besides I daily meet there some whom I know and am glad to see.

March 18th. I have been repeatedly interrupted in writing this letter, but I am determined at last to make an effort to finish it. The new Ministry is certainly of very peculiar composition: the law appointments are good, and the Admiralty, at least in the opinion of the Napiers, much better than the last; but otherwise, the whole seems to rest on Lord Derby himself and Disraeli.

I quite agree with you in disliking especially the way in which Lord Derby speaks of the Church in reference to education; in fact this is to me by far the most objectionable part of his programme. As to the Free-trade question, I find that the most of the people I meet, think that Lord Derby means to throw Protection overboard; but this I fancy he can hardly do without throwing most of his sub-

1852. ordinates overboard too, as they are too deeply committed on the question. Without being myself a very warm Free-trader, I look upon the question as fairly settled, and I think it would be exceedingly injudicious and mischievous to re-open it and to try to reverse the decision. The decision may have been unfavourable to the farmers, but they were beginning to accustom themselves to it, and to make their arrangements and agreements accordingly, and would be much better even for them to acquiesce in a *fait accompli*, than to revive an irritating and dangerous discussion. If this question were once out of the way, I do not know what should prevent Lord Derby from coalescing with the Peelites, and thus forming a much better Ministry than the present ; though in one point of view I should regret this, as it would drive the Russellite Whigs into the arms of the Manchester party. I think the elections in France afford an instructive lesson as to the practical effects of universal suffrage and vote by ballot, with an ignorant and priest-ridden population.

It is reported that Louis Napoleon is certainly to be proclaimed Emperor at the great review on the 1st of May, and indeed when he possesses the reality of absolute power in such perfection, it is absurd to affect shyness about the title. The success of that usurpation is a most curious phenomenon, hardly to be paralleled, I think, since the fall of the Roman Empire. Pray give my love to Mrs. Horner and Leonora.

Ever your affectionate Son-in-law,
C. J. F. BUNBURY.

To Lady Bunbury.

4, The Grove. Highgate,
April 29th 1852.

My Dear Emily,

I am sorry to say that I cannot give you 1852. much assistance about the Jungermannias, at this distance. If you collect specimens, I may perhaps be able to name them, or at least to help you to name them, when we meet, but I have studied them less than the Mosses *proper*. Neither do I know of any book which will help you much, except Sir William Hooker's splendid work, which I could have lent you, if we had thought of it in time, but it is too bulky to send so far; it is a quarto with beautiful coloured plates of all the British species.

I remember that the Jungermannias are particularly abundant and beautiful at Abergwynant, and I collected a great quantity there in '46, which I have never named or arranged.

We had some very fine days at Ventnor, really spring, and the day I was at Alum Bay was brilliant. I spent a very interesting and agreeable day in geologizing there, and another almost equally interesting (in spite of bad weather) at Compton Bay and Brook Point. We had a very pleasant drive from Freshwater to Ryde; and between Newport and Ryde, at the edge of a coppice, by the roadside, Fanny's sharp eyes detected a very rare plant, Pulmonaria angustifolia. It was first discovered in that spot by Dawson Turner, many years ago, and has been found in only one or two other places in Britain.

1852.   The Lyells are in great force, and in high spirits
about their mission to America.   Edward, whom
I saw yesterday, is looking well.   Sir Charles Lemon
seems to have recovered from his attack ; at least
he is about again in society much as usual, and
Mrs. Horner, who has seen him, says that he
shows no sign of ill-health.

I am much obliged to my father for his letter,
and will write to him very soon.

Pray give my love to him and Cecilia and Henry,
and

Believe me ever your affectionate Step-son,
C. J. F. BUNBURY.

To Mrs. Henry Lyell.

Mildenhall,
May 10th, 1852.

My Dear Katharine,

I have not written to you for some time,
but I have thought of you very often, and have read
with great delight and interest all your charming
letters from India.   There is a freshness, and clearness, and spirit, about all your accounts of your
journeyings, and of the people you see, and of all
you observe and do, which makes them quite
delightful, and I assure you I am very much
interested too in all your accounts of dear little
Leonard and his new brother, although, not having
any personal acquaintance with *Babyology*, I cannot
so fully enter into the annals of the nursery as

some others can. It made me very happy to hear 1852.
of your safety and welfare; and I do most earnestly
trust that, before many years are over, we may
have the comfort of seeing you and your Husband
and *all* your little ones safe again in old England.
I have felt half inclined to envy you your voyage
up the Ganges, and the opportunity of seeing such a
remarkable country; but then I know that I
should get most excessively impatient of the restric-
tions which the nature of the climate imposes on all
who do not wish to kill themselves, and which
are particularly annoying to a naturalist. There
certainly seems, by all accounts, to be something
peculiarly mischievous and deadly in the sun of
India; even at Rio de Janeiro, which is much
nearer to the equator than Meerut, I could go out at
all times (when it was not raining) without positive
risk,—although I must own that in the hot season
it was not quite pleasant. Even in Ceylon the
heat does not appear to be so formidable as in the
North of India, for this and the other reasons you
have mentioned, I shall not be at all surprised if
you are able to collect little or nothing in India,
unless indeed you are able to go to the
hills.

Your Cape collection was really a surprising one,
for the short time in which it was made, and
considering too your delicate health.

Joseph Hooker made many inquiries about you,
and was much interested by the accounts I gave
him of you. I enjoyed much our stay in London,
and at Highgate, and absolutely *revelled* in the

1852. Athenæum, into which I hae been elected in February, and which I found an immense comfort and resource. Such a library! I spent my days studying and extracting from valuable foreign botanical works, known to me before only by name : above all, Von Martius's splendid works on the Botany of Brazil. It is confessedly the best club library in London, and besides, Hooker says it is the only club he knows which is *quiet* enough for one really to enjoy the library. Moreover, I genèrally met there some one or other whom I was glad to see.

We spent three delightful days at Waverley, and then came home, and here we shall probably be fixed for the next six months at least.

Poor Mrs. Power is staying with us at present* and very pleasant she is,—cheerful, quiet, and gentle, and full of good nature and kind feeling.

I have been reading the Life of Lord Jeffrey, and reading it with much pleasure and interest ; but to those who knew him well, and who were familiar with Edinburgh, it must of course be much more interesting than to me ; and I daresay you will enjoy it very much indeed.

Then, in my own more especial way, I have examined, since I came home, a collection of fossil plants which Daniel Sharpe had received from Portugal,—quite a new field for fossil botany, and have written a short account of them for the said Daniel. I am creeping on with my big book on

---

* Mrs. Power had lost her Husband, Major Power, in December, 1851. She was Mrs. Bunbury's Aunt.

fossil plants, and I am working at my Uncle Fox's 1852.
collections from South Brazil, and Buenos Ayres,
on which I intend to write a paper for the
Linnean ; they take a long time to go through,
on account of the enormous number of duplicates.
I am, moreover, rearranging my whole collection of
Mosses, according to Bridel.

Ever your very affectionate Brother,

C. J. F. BUNBURY.

To Lady Lyell.

Mildenhall,
11th July, 1852.

My Dear Mary,

Though Fanny is writing you an epistolary
volume, and will leave me nothing to say, I must
write a few lines to congratulate you with all my
heart and soul on the twentieth anniversary of your
wedding day. I do not know two people more
entitled to congratulations on such an occasion.
That every blessing may attend you both, and that
you may both grow old together in the enjoyment of
as much happiness as is compatible with human
nature is my most hearty wish.

We have had glorious summer weather, worthy
of Italy : this is now the 9th day of it. In London
I should think it must have been oppressive, but
here, with the shade and green and flowers, it has
been delicious.

1852.   Fanny has twice driven me out in her little ricketty carriage, and we have not broken down, which I think is a wonder.

The heat agrees very well with me, and I am working hard both at my fossil Ferns, and at my paper on Buenos Ayres vegetation.

You will have heard how the Bury Tories succeeded in defeating Edward by force of money and beer. I am afraid it is a great disappointment to him, though not so much so as a similar failure would be to some men, because he has never made political life his *sole* object. With a few exceptions the elections are going on triumphantly for the " Liberals." I am only afraid we shall have too Radical a House of Commons—a fear in which I know you will not sympathise. However, I shall not be sorry to see the present Ministry out, for Lord Derby's high church leaning, and Lord Malmesbury's truckling to Austria, have disgusted me with them. The Irish elections seem likely to be terrible.

With much love to Lyell,

Ever yours, most affectionately,

C. J. F. BUNBURY.

JOURNAL.

September 9th.

Arrived at Edinburgh about 2½ p.m.   Douglas's 1852. Hotel.

September 11th.

Move into lodgings, 17, Melville street.   Spent 2½ hours in Botanic Garden.

September 12th.

Very cold.   To St. John's Church, and heard an excellent sermon from Dean Ramsay.   Drove to Bonaly and spent the afternoon there.   Distant view of some of the Highland mountains from Bonaly.

September 13th.

Very fine day.   To Lauriston Castle, and saw Lord and Mrs. Rutherford, very friendly.   Beautiful views of the Firth, and the hills of Fifeshire, from the terrace walk at Lauriston.   Good collection of Conifers.   Return along the sea-shore, by Granton on the chain pier, but found no sea-weeds worth picking up.

September 14th.

Fine, but cold wind.   Dine at Lauriston Castle ; Lord Rutherford extremely agreeable.

1852.   A lovely day.   Drove with Fanny down to
Granton pier, then walked for some time along the
shore, looking for sea-weeds, but with little success.
The   common   Fuci,   Laminaria   digitata   and
saccharina, Ulva latissima, Porphyra vulgaris,
Enteromorpha compressa in profusion, but I found
hardly anything at all uncommon, except a fragment
of Delesseria sinuosa.   Came home at half-past
four, and afterwards walked under the Dean Bridge
(a noble structure) and some way down the water
of Leith, as far as Stock Bridge.

September 17th

Mr. and Mrs. Horner and Joanna arrived, and
I went with them to Lauriston Castle.   They
dined with us.

September 18th.

To the Botanic Garden with Joanna.   We all
dined at Lauriston ;   met Sir W. Gibson Craig
and Mr. Andrews,

September 19th.

Went to a " Free Kirk " Church to hear Mr.
Guthrie preach.   Sermon very eloquent, but I do
not like the Presbyterian service.

September 27th.

A fine day.   Still confined to the house.   Read
some more of Henfry, and began reading Brace's
" Hungary in 1851."   Curious account of the im-

mense plains of Hungary, resembling the Pampas of 1852.
South America; no wood or stone; houses all built
of mud and reeds, and fields fenced with reeds.
Interesting account also of Kossuth and Gorgey.

[NOTE.—They had intended to pay various visits in
Scotland and to make a tour in the Highlands, but
his Wife caught cold and had a dangerous illness,
so that all these plans were broken up and they
were detained in Edinburgh until the middle of
November].

LETTERS.

17 Melville Street,
October 12th, 1852.

My Dear Mr. Horner,

We both feel very much your unselfish
kindness in consenting to part with Mrs. Horner
and Joanna, and we are very sorry indeed, that you
should be left quite alone, which we did not contemplate when Mrs. Horner first proposed to come.
But if Fanny goes on as well as she now seems to
promise, I hope they will soon be able to rejoin you.

Poor Mrs. Rutherfurd's sudden death is a very
sad event, and she will be much missed.

I have been reading " Boswell's Tour to the
Hebrides," partly aloud to Fanny, which to me is
a singularly entertaining book.

Believe me ever,

Your affectionate Son-in-law,

C. J. F. BUNBURY.

3 H

To Lady Bunbury.

127 George Street,
Edinburgh.
October 24th.

My Dear Emily,

Your very kind and pleasant letter of the
12th was very welcome, and I was very much
obliged to you for it, though I have left it nearly a
fortnight unanswered. I had an unusual quantity of
letter writing to do while dear Fanny was ill, and
since she has been well enough to write her own
letters, I have relapsed into my usual ways. I am
very thankful to be able to say that she is now quite
convalescent, indeed I trust nearly well, except that
she has not quite picked up her strength yet; and
she is looking much better than one could have
expected so soon after such a severe illness. I do
earnestly hope now that we shall soon be able to
move to the South, and I am the more anxious to
move while the weather *is* tolerable, because I believe
with Henry that we shall have a severe winter.

Fanny was so well that she was able yesterday to
accompany us to Roslin, and she does not seem
at all the worse for the excursion. She did not of
course walk much, but she was able to see the chapel,
which is certainly exceedingly well worth seeing; I
was very much struck both with the beauty and
the singularity of it; the profusion of delicate and
elaborate sculpture is wonderful. The glen was
still in great beauty, with the rich and varied tints
of autumn on the foliage, though we wanted sunshine to see it to full advantage : and indeed we
were a *little* too late,—too many of the leaves had

fallen. Altogether we had a very pleasant excursion 1852.
(we four, that is, Fanny and her mother and sister
and I), and I am very glad to have seen Roslin, of
which I had long had a romantic image from
Scott's poetry. And it was a great satisfaction to
find that Fanny could go through the undertaking
without any injury.

I know you must have felt poor Mrs. Rutherfurd's
death very much; though so sudden, she seems
to have been quite prepared for it herself, but it
must have been a great shock to all her friends;
she was such a gentle, kind, amiable person.

I quite agree with you about "Uncle Tom's
Cabin," such monsters as Legree are exceptional,
and prove nothing. Still it is, I think, a work of
real genius as well as of most excellent feeling, and
I hope it may do good.

How thoroughly natural and well sustained is the
character of Marie! and such a character, being
in itself so common-place, shows the real working of
the system far better than any monster. One effect
the book has had upon me, certainly, is to con-
vince me of the necessity of our *keeping* Canada
most tenaciously, which I was before not satisfied of.

My reading here has certainly not been systematic,
but latterly, since Fanny has got so much better, I
have been able to get to work again at writing
for my *big book* of fossil plants, though I cannot help
thinking that said book is something like poor
Dr. Strong's Dictionary. By the way I have read
David Copperfield for the first time, and with great
delight. 3 H 2

1852.   The present state of France is curiously like that
of Rome under the emperors, and it is quite
clear that if Louis Nap wished to have altars raised
to him, there would be plenty willing to do it.
As long as the army and the priests continue to pull
together, his power seems pretty secure, — and if
he were overthrown, the reaction would be so
tremendous, that I can understand how even
respectable Frenchmen may wish for the main-
tenance of his power, rather than the chaos that
would follow its subversion.

Pray give my love to my Father and Cecilia, as
well as to Henry, and believe me ever,

Your affectionate Step-son,

C. J. F. BUNBURY.

JOURNAL.

October 25th.

Fine day.   Joanna and I walked round the foot
of Salisbury Crags, starting from Holyrood and
returning by St. Leonard's.   In our way back went
to see the Grey Friar's church-yard, where Joanna
showed me the monument (raised in 1706) to the
Covenanters slaughtered in the time of persecution,
from 1661 to 1688 ; the tomb of Robertson ; that of
Blair ; that of Maclaurin the mathematician, with a
remarkably beautiful Latin epitaph ; and that of the
great Scottish lawyer and scholar, Sir George
Mackenzie.   It is an interesting cemetery.

October 29th. 1852.

Went by appointment to see Mr. Bryson's collection of fossil woods; which I found very beautiful and remarkably interesting. Mr. Bryson, a watchmaker, is a very intelligent man with great knowledge of the subject and much simplicity of manners : I like him much.

LETTER.

127 George Street,
Edinburgh,
October 30th, 1852.

My Dear Lyell,

You seem to be enjoying mightily your tour in America, and I have been much interested by your letters, especially by that long one which you wrote to Mr. Horner. I hope you will continue to enjoy yourselves, and that we shall see you safe and well at Christmas.

The death of the great Duke has been the chief public event since you went away, and the general expression of feeling on the occasion was very satisfactory. In France, the Empire is not yet come, but is undoubtedly close at hand. Whatever we may think of their taste, it really looks as if the great majority of the French approve of Louis Napoleon, and we certainly have no right to interfere with their choice.

With much love to Mary,

Ever yours affectionately,

C. J. F. BUNBURY.

JOURNAL.

1852.   Mrs. Horner and Joanna left us to return to
Highgate.

Went to hear Macaulay address his constituents,
Got a pretty good place, and heard him with
interest and satisfaction, though I could not catch
all he said.   His delivery I thought at first, mouthing
and declamatory, but improved afterwards; I do not
however think it very distinct.   The speech a good
one, but on the whole scarcely so brilliant as I
expected.   He was very well received.

Notes on some Books he read while in Edinburgh.

Finished Brace's Hungary ;—an instructive book.
The detestable system of Austrian government since
'49.   The great things done by Kossuth in organizing
the finances and providing for the defence of the
country.   I now understand much better than ever
before, on what his fame rests, independently of his
eloquence.   His influence over the Hungarians
seems to be wonderful.

Finished " David Copperfield," certainly one
of the most delightful of Dickens's works.   The
chapter on the model prison is a true and sound

and forcible piece of satire. But there is something far better than satire or wit in the book ; such true pathos, such exquisite touches of feeling, such a fine and noble moral sense. Agnes Peggoty and her brother Traddles, and Betsy Trotwood, are ·all delightful.

Finished "The Subaltern," a remarkably pleasant and well-written memoir of military life. Read Cicero's "De Senectute." This is the third time I have read it. I was first made to read it by my tutor when I was about 14, when I did not relish or appreciate it at all. I never looked at it again till the winter we spent at Edinburgh, 1849-50, when I read it with great delight and admiration, and these have not been in the least diminished by this last reading. The language and sentiments are alike beautiful. How well he speaks of the consolations of old age ; of the enjoyments of a studious and tranquil old age, and of one spent in rural pursuits and the observations of nature ; of the happiness of looking back on a well-spent life ; How noble are all his reflections upon death ; and surely no one, Pagan or Christian has ever treated more beautifully of the immortality of the soul !

———

During this long and tedious time they spent in Edinburgh, he amused himself by walks in the beautiful neighbourhood, besides a few longer excursions : Arthur's Seat, Duddingston Loch, Lauriston, and the Botanic Garden, were almost daily visited.

1852.   In the middle of November they came to London and stayed in the house, 11, Harley Street, lent to them by the Charles Lyells, who were in America.

November 15th.

To the Athenæum, where I saw Douglas Galton, Robert Brown, Dr. Fitton, Lord Colborne, Lord Monteagle, and George Jones. Talked principally about the Duke's funeral and the floods. Some talk with Robert Brown about fossil woods. He looks feeble and broken. Studied "Boryde St. Vincent on Algæ" (in the Voyage de la Coquille) and made some notes. Read some of "Evelyn's Memoirs," and looked into a French translation of Athenæus.

November 16th.

At the Athenæum most part of the day; saw Monckton Milnes, Sir E. Ryan, G. Jones, and Mr. Stokes. Called on Mr. Donne at the London Library, and had much pleasant talk with him. Studied Von Martius's magnificent work, "The Nova Genera." Read two papers by Auguste St. Hilaire, in the 12th volume of the "Memoirs du Museum," one on "Sauvagesia and its Allies," the other (very curious) on the "Poisonous Plants of Southern Brazil. Read also "The Life of Sir Joseph Banks" in Brougham's "Lives of Literary Men of the Age of George III.;" it is interesting. Edward drank tea with us. Everybody is talking about the Duke's funeral; there is hardly any other topic of conversation.

November 17th. 1852.

Called at 1, Hobart Place, and saw Minnie and Lady Napier, and had a very pleasant talk with them. They are going to St. Paul's to see the funeral. They say there is a report current in Hampshire that the French are going to make their invasion to-morrow. Then to Cadogan Place, and saw the Richard Napiers. Then to Somerset House, to the Council of the Geological; the meeting lasted nearly two hours; much discussion about the suggested removal of this, with other scientific societies to a new site, and about the danger of amalgamation. Edward Forbes proposed by Hopkins as the next President, with the hearty approbation of all. Walked back to the Athenæum with Hooker, and had some talk with him. Difficult to get along the streets for the crowds and the barriers and scaffoldings. Dined at the Athenæum with Mr. Sam. Smith, of Combe Hurst, a very agreeable man; much pleasant talk. Then to the evening meeting of the Geological Society; a very thin meeting, one of the thinnest I have ever seen. Mr. Bains' important paper on the "Geology of South Africa" was read, or rather a copious abstract of it by Rupert Jones. Murchison and Owen the principal speakers; the former on the Palæozoic fossils, the latter on the gigantic reptilian remains.

———

November 18th.

Very fine and rather cold. The noise and bustle great, even in this quarter, long before daylight.

1852. By and bye, came Mr. and Mrs. Horner and Leonora and Edward, who had all seen the procession from the Athenæum. All had gone on well so far.

_____

Dined with the Horners at Highgate. Took Susan and Mr. Donne up thither in the fly with me. A very pleasant party. The Bishop of Manchester, Mr. Lingen, Donne, Lady Bell, and Mr. Shaw, in addition to the Horners themselves. Very good talk during dinner. The Bishop evidently a man of powerful mind as well as of much knowledge, his conversation rich, clear and vigorous. In the evening talked principally with Leonora about botany and various other things. Brought Mr. Donne back to town and had very pleasant literary talk with him.

_____

Met my Uncle William at the Athenæum and had much talk with him. Unlike most people that I have met, he does not think a French invasion probable, being of opinion that it would not suit the objects, or the policy of the President Emperor; but he agrees with Sir Charles that we are not in a fit state of defence if such a thing should be attempted. He finds great fault with the appointment of Lord Hardinge to be Commander-in-Chief, instead of Lord Fitzroy Somerset, says that it was quite contrary to the known wishes of the Duke, and that all the army is indignant at it.—Dined with the Douglas Galtons in Chester Street; no one else

of the party. After dinner, Douglas Galton took me 1852.
to the meeting of the Geographical Society (in the
lecture room of the Royal Institution) where I heard
Captain Englefield read an account of his voyage in
search of Sir John Franklin. His paper was sadly
deficient in simplicity and sobriety of style ; what
he said viva voce in the discussion was better. Sir
John Ross, M. Petermann, and Captain Penny, took
part in the discussion which turned mainly on the
question of the communication of Baffin's Bay with
the Polar basin.

November 26th.

Returned to Barton.

November 27th.

A most beautiful day. Spent all the morning
in walking about the arboretum and grounds, first
by myself, then with my Father, and had much
comfortable talk with him. Walked down to "The
Cottage," and saw Sally and the children. At
dinner, Lady Cullum and Patrick Blake, both
entertaining and pleasant. The militia here well
spoken of.

November 29th.

Aunt Caroline,* Catty and Norah, William Napier,
John Alcock† and Edward arrived. Minnie unfortunately not able to come. A grand jollification
in the evening in Fanny's room.

November 30th.

Henry's marriage with Cecilia. *Sit felix faustumque.*

---

* Lady Napier, his Mother's Sister.　　　† Afterwards John Herbert.

1852. The weather sadly unfavourable, raining violently
and very cold. Cissy looked extremely well in her
bridal dress. Her bridesmaids, Catty and Norah
and little Louisa. Patrick Blake, "best man."
The Vicar performed the service very well. The
wedding breakfast very successful, and the Vicar
made a neat speech with a good deal of feeling, in
proposing the health of the bride and bridegroom.
In the evening a very merry party in the little
library. Fanny in very good spirits this day.

December 3rd.

A very fine day. My Aunt and Catty left Barton,
and in the afternoon we returned to our home,
which we found in nice order, looking very cheerful
and comfortable ; the dogs in fine condition. I
have enjoyed this visit to Barton, and especially the
company of my good kind Aunt and her daughters.

They say that Lord Derby has given Lord
Palmerston the name of *Don Pacifico,*—very happily
applied, certainly. Other witticisms of Lord Derby
are told. When his Ministry was first formed by
the assistance of D'Israeli, he called it "Benjamin's
mess." At the beginning of the dispute with
America about the codfishing, Lord Malmesbury
avowed that he was very ignorant on the subject,
and asked Lord Derby how he could best obtain
information : he advised him to take *cod-liver oil.*

A very judicious, well-drawn and discriminating
character of Mr. Hasted in the "Bury Post,"
probably written by Arthur Hervey.

December 4th. 1852.

Busy in arranging matters in my museum, and getting my materials into working order. Heard of the defeat of the Whigs at Bury. In the evening, began reading to Fanny, Prescott's History of Peru.

December 5th.

Sunday. Spent the morning in looking over various parts of my botanical collections. Went to afternoon Church. Read to the servants a sermon of Arthur Hervey's in the evening.

December 6th.

Began the ninth book of the Iliad (having read the first eight about two years ago).

December 7th.

Read four chapters of the History of Chemistry, in Whewell's "History of the Inductive Sciences."

December 8th.

In the evening read a little of Prescott to Fanny, and read to her also, in illustration, Humboldt's description of the wonderful Peruvian roads.

December 9th.

Spent part of the morning in the garden, and ordered the planting of several plants of Berberis aquifolium. The Coronilla glauca now in full blossom in the open air, and some of the Fuchsias, the Salvia fulgens, a Penstemon and the yellow China Rose, partially in flower. The Chimonanthus in bud. Read the eighth chapter of the fourteenth

1852. book of Whewell, which treats of the Atomic theory, or (as it is more safely called), the law of definite and multiple proportions. This law seems to have been partly discovered before the time of Dalton, but had attracted no attention, and he was the first who developed and established it.

December 10th.

Board of Guardians; much business, more than thirty applications for relief; many able-bodied men out of work. Very little business at the Petty Sessions

December 11th.

Meeting of the committee of the Coal and Medical Clubs. Read to Fanny the remainder of the 3rd and part of the 4th chapter of Prescott's Peru. If all that is told of the Peruvian system of Government be true, it was certainly the most extraordinary that ever actually existed, or rivalled only by that of the Jesuits in Paraguay;—and the latter was on a much narrower scale. It is hardly conceivable that the whole population of a large country could for many generations have been contented under the most searching of despotism which left them no freedom of action or of will in any incident of their lives, and which reduced them in fact to mere machines.

December 14th.

Received the joyful news of the Lyells safe arrival in England.

Board of Guardians from 10 o'clock till near 1 ; much business.

———

December 20th.

To Bury. Had a long chat with that most good-humoured and cheerful person Lady Walsham. Heard that Lord Aberdeen has undertaken to form a new Ministry.

———

December 24th.

Read a good deal of the 2nd volume of Humboldt's Cosmos. Read to Fanny, Milton's fine " Hymn on the Nativity," and thirty pages of Prescott.

———

December 27th.

Leonora and Joanna arrived, and we passed a very pleasant chatty evening. Read the remainder of Whewell's chapter on Linnaeus.

———

December 28th.

A beautiful day. Took an hour's walk ,with Leonora ; the first walk I have taken, beyond our garden for more than a fortnight. Joanna sang to us in the evening. Read the 6th chapter of Whewell, which concludes his history of systematic botany. This history is very well done, and gives, though very concisely, a clear and just account of the different steps in the progress of the science, and of the difficulties overcome.

———

1852.   December 29th.

Louisa and Harry arrived.

December 30th.

Took a walk with Leonora and the two children, and saw two Fir trees which had been blown down in the Folly by the gale of the 27th.   Mr. and Mrs. Horner and Susan arrived.   A very pleasant social evening.   Looked over a handsome illustrated work on Sicily, which Mrs. Horner had brought, the views were very good.   Mr. Horner much pleased with the composition of the new Ministry, as I am too.

December 31st.

Friday   Board of Guardians ;   17 applications ; 6 able-bodied people out of work ; the rest cases of sickness, chiefly influenza and rheumatism.   Wrote to my Father, and just as I had finished the letter, Henry arrived, and stayed to luncheon.   The Charles Lyells came soon after, and Edward a little before dinner.   Games and merriment in the evening.   We all (except Fanny) sat up till after midmight and welcomed the new year.

So ends '52.   God grant that the coming year may be well spent, and that my beloved wife may be in better health at the end of it than at the beginning.

JOURNAL.

Received a very agreeable letter from Katharine, 1853. from Umritsir. Read in the newspapers the account of the French Emperor's marriage.

February 2nd.

A beautiful day. Occupied all the morning with South-American Verbenacæ and Labiatæ, except some time that I spent very pleasantly in showing part of my collection (the Asclepiadeæ principally) to Fanny.

February 4th.

My 44th birthday, Received several kind letters and presents from our friends. All the morning very disagreeably spent in the drudgery of the Board and the Bench. In the afternoon my Father and Lady Bunbury arrived. They brought me a parcel of specimens of minerals and fossils (some of them curious), together with a letter from Henry. The rest of the day passed in very pleasant conversation.

February 5th.

I read some of "Francis Horner's Life," a book I always recur to with great pleasure.

LETTER.

Mildenhall,
February 6th.

My Dear Mrs. Horner,

I thank you and Mr. Horner very heartily for your kind expressions and good wishes, as well

1853. as for your pretty gift.   I did not know the little
book* before, except by name, but it is very
welcome. Fly-fishing indeed, is a subject of which
I am profoundly ignorant, whether as to theory or
practice, nor am I very likely to take up the study,
but the book seems to contain a good deal of curious
matter bearing upon natural history, and Sir Hum-
phrey Davy is so agreeable a writer, that everything of
his is worth reading.   The vignettes are very pretty.
But I assure you I value your expressions of regard
and kindness far more than any presents, and your
affectionate note, as well as those from the dear
Sisters, give me very great pleasure.   I have little to
tell, except that my Father's and Lady Bunbury's
visit was very pleasant in every way, and we had
nothing to regret but that it was so short.   They
seemed well and in very good spirits, and were
extremely agreeable.   To-morrow we expect Mr.
Donne and the Arthur Herveys, and on Tuesday
my brother Henry.

Believe me ever your very affectionate Son-in-law,

C. J. F. BUNBURY.

JOURNAL.

February 7th.

The Arthur Herveys and Mr. Donne arrived;
very pleasant conversation.   Mr. Donne's lecture
on Alfred the Great full of information and in
every way very instructive.

* " Davy's Salmonia."

Mr. Donne left us after breakfast; the Arthur 1853.
Herveys stayed till the evening, and we had a very
pleasant day with them. Arthur Hervey is a great
favourite with me, and equally so with Fanny. He
has a truly beautiful mind, peculiarly pure, refined,
candid and liberal; full of benevolence and Christian
charity, and of enlightened love of truth, and a
considerable variety of knowledge and accomplishments. I always find it does me good to be in his
company. His wife is worthy of him.

February 13th. (Sunday).

Read to Fanny an admirable sermon by Channing
on "Self-denial." By the way, self-*denial* is not a
very accurate or intelligible name for that virtue,
but rather calculated to convey false ideas; self-
restraint is a much more correct term.

February 16th.

In my walk saw several hooded crows.

LETTER.

Mildenhall,
February 17th, 1853.

My Dear Susan,

I owe you many thanks for your kind note
and good wishes on my birthday. We have been
busy looking over our library, and putting up a
quantity of books and pamphlets to be bound. We
have both been reading "Moore's Life." It is

1853. very pleasant reading; the diary in particular very entertaining, and giving me an extremely pleasant impression of Moore himself, and of Lords Lansdowne and Holland. We are likewise going on with Prescott. I was much grieved to see in the newspapers the account of the insane attempts at insurrection at Milan. Such attempts under present circumstances seem mere madness (like poor Emmet's rising at Dublin) and can lead to nothing but fresh oppression, slaughter and misery. If Austria were actually at war with Turkey (which seems likely enough to be the case before long) or with France, which is not quite out of the question either, there might be some chance for the poor Italian patriots, but the rashness of the present outbreak is inconceivable.

With much love to Mr. and Mrs. Horner, and wishing you all well out of the snow,

I am ever your affectionate Brother.

C. F. J. BUNBURY.

JOURNAL.

February 19th.

Took a walk in the snow with the two dogs; the drifts pretty deep in some places. The fir plantations in this weather, with the snow lying heavy on their branches, and the snow beneath, have a wild, gloomy, almost romantic look.

February 20th, Sunday.

Read to Fanny Channing's second Sermon on Self-denial : excellent. Read to the servants in the evening a good sermon of Arthur Hervey on Christian Humility. Finished the first volume of "Moore's Life." I still think that the collection of letters much required weeding. Lord Moira's behaviour to Moore certainly shabby; only an additional instance of what men of genius have to expect who look for the patronage of the great.

By the way, Lady Bunbury, when she was here, told me a good deal about Lord Moira, whom she knew well. His stately and graceful, though rather formal courtesy; his lavish generosity; his boundless ostentation and extravagance, of which one comical instance was, that when he was going out to India, his children being very young, he provided them with *fifty black dolls*, in order to accustom them beforehand to the sight of dark faces. He was very fond of botany, and told Lady Bunbury that he had owed to it some of the most delightful hours of his life.

February 23rd.

Sally and her children arrived late.

February 24th.

Read Lord John Russell's preface to the Memoirs of Moore, and skimmed over parts of "Lallah Rookh" which I have not read for many years. How exquisitely beautiful are those lines of Moore which Lord John quotes at the end of his preface :—" This world is but a fleeting show," &c.

1853.   Sally and the dear children left us.

———

A visit from Mrs. and Miss Bucke, whom I had not seen since September.

———

Read the eighth and last chapters of the eighteenth book of "Whewell's Inductive Sciences."

———

A mild day ; snow almost all gone.   Preparing for departure.   Looked over and put by a variety of articles in my museum.

———

London.   Henslow in the same carriage with us as far as Cambridge ; had much talk with him ; *inter alia,* on the Mistletoe ; he concurs in my notion of the confusion between our common Mistletoe and the Loranthus Europæus, being the explanation of the contradictions and perplexities in Pliny's account.

Two of the Miss Waddingtons our companions to London.   Saw Mary, Leonora and Joanna.

———

Weather warm and wet.   Saw my Father and Lady Bunbury, Henry and Cissy.   Looked into the British Institution, but saw no pictures that attracted me particularly.   At the Athenæum consulted the Flora Antarctica and other books, and

made some extracts. Dined and spent the evening 1853.
with my Father.

———————

March 9th.

Very fine warm day. To the National Gallery,
and saw the two pictures by Turner, — " The
Building of Carthage,"—and " The Sunrise in a
Mist," which are newly placed there. They are
really glorious pictures ; I had no idea, before I saw
them, of Turner's merits. They certainly do not
suffer by a comparison with Claude. — Then to
the Athenæum to get my luncheon and to look
at the newspapers,—then to the Geological Society,
and on my way looked in at Sowerby's, where
Dawson Turner's library is selling by auction.

The Council of the Geological Society very fully
attended.—Then to Harley Street, and had a
pleasant chat with Mary and the girls. Heard
of the death of Von Buch.

Dined at the Geological Society's Club ; a very
full attendance. Murchison in the chair. Sat
between Lyell and Daniel Sharpe ; Austen opposite ;
Moore on the other side of Lyell.

Meeting of the Geological Society. A curious
paper by Mr. Dawson, on the Albert Coal Mine in
New Brunswick ; followed by a very good discussion.

The substance from the Albert Mine seems to be
nearly pure bitumen, and is said to melt by heat into
a kind of naphtha or petroleum ; it is intensely
black, very brittle, light, with a peculiarly brilliant
lustre, and without any crystalline structure. Its

1853. commercial value is great, on account of the **very** large proportion of gas which it yields. Reference was made in the discussion to the very similar litigation which is going on in Scotland about the so-called "Boghead Coal," which is a highly bituminous schist, worked for gas with very great profit. In this case also, the question is whether the substance worked is or is not comprehended under the general name of coal. Edward Forbes, our president, closed this part of the discussion by saying that the legal question had better be referred to "Coke."

March 10th.

Another very fine day. From London by the Brighton and South Coast Railway ·to Havant, where the William Napiers met us, and took us up to their house at West Leigh.

LETTERS.

Ventnor,
Isle of Wight,
March 29th, 1853.

My Dear Katharine,

Your letter of December 20, which I received on the 1st of February, was a very agreeable one, and ought not to have remained so long unanswered. I know that the rest of the family keep you regularly informed of all that happens, so I need not tell you anything about Cecilia's marriage, nor dilate on our delightful Christmas party, which,

however was a thing never to be forgotten; from the 1853.
arrivalof the *first detachment*, in the shape of Leonora
and Joanna, to their final departure, was altogether
three weeks, and I do not think I ever spent three
pleasanter weeks. Joseph Hooker spent several
days with us, and was very pleasant ; Fanny, who I
think did not know him much before, was very
much pleased with him : he certainly is a remarkably
agreeable man,—so much cheerfulness, vivacity,
frankness, and activity of mind, immense knowledge,
and so much readiness to impart it, and to help
others in their pursuits. I had a great deal of
satisfactory talk with him, and learned much.
Henslow also was with us part of the time, and he
too is a most valuable man ; the Father-in-law and
Son-in-law are worthy of each other. Hooker is
very busy with his great Indian collections, and he and
his friend Dr. Thomson (who travelled in Thibet,
and the western Himalaya) are working together at
a great work, no less than a general Flora of
India, for which they have immense materials,
having the use of the collections made by Roxburgh,
Wallich, Boyle, and Falconer, as well as their own. It
must be a work of time, as the Flora will probably
contain more species than Linnæus knew of as existing in the whole world ! It is not, however, to be a
splendid work like the Flora Antarctica, which is
really an incomparable book, I study it again and
again whenever I am in London, and find it an
inexhaustible mine of botanical knowledge.

I was interested by your account of the Asclepias gigantea, the castor-oil plant (which I

1853. have seen growing magnificently in Brazil, and the prickly grass. I daresay this last is a *Cenchrus*; there is a species of that genus, C. tribuloides, at Rio de Janeiro, which is a very great nuisance, just as you describe, with the sharp tenacious prickles of its husks.

Since I began this letter, your last have been forwarded to us, containing the account of the inspection of the troops, which has interested us much. I am very glad the 43rd was so much commended. I should like very much to see Harry Lyell at the head of his regiment; I have no doubt he looks extremely well and commands extremely well.

I do not remember whether I mentioned in my former letter, that last summer I wrote a paper of some length on the botany of Buenos Ayres, chiefly from the materials afforded by my Uncle Fox's collections; this has now been read at the Linnean Society, and will, I hope, be printed; and if, as I expect, they allow me any separate copies, you shall have one : Hooker read it while he was with us at Mildenhall, and approved it much, which was a great satisfaction to me.

I am now expecting a grand addition to my cryptogamic collections. My friend Harvey, (the great authority on sea-weeds), to whom I sent a parcel of my Brazilian and Buenos Ayrean duplicates, is sending me a set of exotic Algæ, from North America, New Zealand, the Antarctic Ocean, and other distant regions, which will have additional value as being named by him. I hope to show them to you some of these days.

We arrived at this pretty place on Saturday last, 1853. Easter eve, and are as much pleased with it as we were last year. I hardly know a more agreeable spot on the English Coast, and the climate agrees remarkably well with both of us. We hear, however, that the winter, (that is, the latter part of it) has been very severe, even here, and that the snow has been four or five feet deep along the Undercliff, which I should suppose was a very uncommon event. At present we have beautiful weather, and are out most part of the day. Fanny I am happy to say is decidedly improving in health.

We spent a fortnight with the William Napiers, and two days with Sir Charles, before we came here : both very pleasant visits. We expect to spend about a month in the Isle of Wight, and then to stay some little time at Highgate, and in Harley Street, before we return home.

The chief topics of late with reference to the political world, have been the chances of a French invasion, and the affairs of India, which have at last been attracting a good deal of attention.

I am afraid it is becoming pretty plain that our government of India has been wofully bad, and it is no slight responsibility, to have the welfare of so many millions of our fellow creatures depending on our measures. It is no doubt a deep and difficult question, *what* will be the best government for India ? but at any rate the attention of the public in this country seems to be now awakened to the subject, and I trust the Ministry will not be

1853. able to get the business hurried over as they seemed disposed to do.

Pray give my love to your Husband, and believe me ever,

Your very affectionate Brother,

March 31*st*.   C. J. F. BUNBURY.

West Leigh,
April 24th.

My Dear Mrs. Horner,

What gay people you are at Highgate ; you seem to have a constant succession of parties abroad or at home. Fanny will certainly not be able, at least it will not be prudent for her to go out either to dinners or evening parties, but I am quite willing to dissipate a little ; though as to the grand London midnight crowds, I am entirely of Mr. Horner's way of thinking. I wonder whether you will think Fanny looking well. She seems well and in good spirits, and able to enjoy herself during the last week we were at Ventnor. She enjoyed the warmth and the pretty country we passed through in the drive from Freshwater to Ryde ; but the day's journey altogether was rather too long, and she was much tired when we arrived here ; the company and cheerful conversation of our friends however did her good, and she has had a good night. I enjoyed our three days at Freshwater, and had some famous geologising. I had a glorious day at Alum Bay, and was charmed both with the scenery and the geology. I collected a few fossils which I will show Joanna. The two remain-

ing days of our stay were very unfavourable, but 1853.
nevertheless I made my way to Brook Point,
examined the interesting geological section in the
cliffs at Compton Bay; saw poor Mantell's fossil
forest (his account of it is rather exaggerated by the
way), and brought back plenty of specimens, so I am
very well satisfied with this second visit to the Isle
of Wight, though I must own that I have had
enough of Ventnor. Yesterday Fanny made a great
botanical discovery; as we were travelling from
Newport to Ryde, her quick eyes detected a very
rare plant, the Pulmonaria angustifolia, which I
had never seen before, and we stopped and secured
specimens of it. This is for Leonora's information.

Mr. Boxall was obliged, by news he received from
London, to leave Freshwater on Thursday morning.
He is an accomplished and agreeable man.

I hope we shall be with you on Tuesday, and I
look forward with great pleasure to our visit to
Highgate. With much love to Mr. Horner and the
sisterhood, believe me,

Your affectionate Son-in-law,
C. J. F. BUNBURY.

JOURNAL.

April 26th.

Arrived at Highgate. Heard that Charles Lyell
is going again to America, as one of the Commis-
sioners from Government to the Exhibition of
Industry at New York; his colleagues, Lord

1853. Ellesmere and Mr. Dilke.   Mary of course going with him.

Went with Mr. Horner and Joanna to an evening party at Edward Forbes's ; met Charles and Mary Lyell, Joseph Hooker, Daniel Sharpe, Murchison, Sir Charles Fellowes, and various others.   Both the Lyells in high spirits about their New York trip. They are to go the 18th May, and do not mean to be away above two months.   Mary in great beauty. Good scientific talk with Forbes, Hooker and Sharpe.   Hooker has been examining my specimens of fossil wood from Natchez, which he has had sliced, and finds its structure very curious, but has not yet been able to determine its nearest recent analogue. He wants to compare it with the wood of Magnolia. It is certainly *not* coniferous.   The fossil wood from Tasmania (*that* with the peculiar fissured appearance) is too highly silicified to show satisfactory structure ; he is inclined to think it coniferous.   He introduced me to Dr. Thomson, who has travelled and botanized so extensively in Thibet and the Himalaya.   I had much talk with him.   Speaking of the Himalayan Pines, he and Hooker both believe Pindrow and Webbiana to be the same species, but Captain Strachey thinks them distinct. Khutrow and Smithiana are absolutely the same thing, not even varieties.

The thoroughly Siberian character of the Thibet vegetation, very remarkable on the Kossyra hills (the climate of which is the wettest known), the forest vegetation of the flanks and valleys is most luxuriant, but the tabular tops, though clothed with

rich and vigorous herbage, are quite destitute of 1853.
trees.

Dr. Thomson (like Hooker) believes the flora of
Brazil to be the richest in the world. In the damper
parts of India, and the Indian Islands, the vegetation
may be equally grand and luxuriant, but is less
varied; the species less local than in Brazil.

April 27th.

Gave Joanna the fossil shells I had collected at
Alum Bay. Took a little walk with Susan. Mrs.
Jamieson (the authoress) came to dinner and stayed
the evening. She is very lively as well as accomplished, and very agreeable.

Mr. Horner approves very much of Gladstone's
Budget.

April 28th.

Met Edward at the Athenæum, and had a good
talk with him; he says there is a strong feeling
at last roused about India, and a very general
opinion against the East India Company; everyone
he has talked with on the subject feels that the
power of the Company must be abolished.

It is thought that the Budget will certainly be
carried even without a dissolution of Parliament.
Edward thinks it of great importance, as calculated
to extinguish for a considerable time the agitation
of the radical "Financial Reformers."

April 30th.

A very pleasant dinner at Highgate; company:—
Mary Lyell, Dr. and Mrs. Hooker, Mr. and Mrs.

1853. Claude Erskine, Babbage and a Mr. Brown from Boston. Mr. Claude Erskine is Son of the Iudian Mr. Erskine (whose conversation I used to enjoy so much during our stay at Edinburgh in the winter of 1849-50) and grandson of Sir James Macintosh. His wife is very pretty and attractive, with vcry sweet, gentle manners, and seems intelligent. I had a good deal of talk with her about Indian trees, flowers, fruit, &c. Had some good talk with Hooker about botanical matters —structure of coniferous wood, fossil fruits, &c.

Sunday, May 1.

A beautiful day. Went to Church with Mrs. Horner. Joseph Hooker came early in the afternoon, and stayed to dinner; and he was very agreeable. He and I, with Mr. and Mrs. Horner, took a walk in the fields, which are just now beautifully green and fresh; I had not expected such thoroughly *rural* scenery so near London.

Talking of the cultivation of cotton in India, Hooker said that the impediments to its success are not merely the want of good roads, the indolence of the natives and their want of care or skill in picking and cleaning the cotton, but that they have already a very good market in China for the inferior kind of cotton which they actually cultivate, and hence have the less inducement to raise a finer kind. He admits that the East India company have been much to blame in neglecting roads, and in other ways, doing so little for the improvement of the country; and thinks it is time that the anom-

alous government of the Company should be done 1853.
away with ; but he says, and I think with reason,
that it is well the government of India was not
directly in the hands of the Crown during the old
times of flourishing corruption and jobbery.

May 2nd.

Took a walk with Leonora ; visited Mr. Yate's
Lauderdale House, and saw his beautiful collection
of tropical plants, especially the Cycads. He has
made a particular study of this family on which he
has read an elaborate paper before the Linnean
Society, and he cultivates a great number of species
some of which are now very fine plants.

Marianne and Tom Lyell (whom I had not seen
since '49), and Miss Parker,* came to dinner.

May 4th.

We went into town to stay a week with the
Lyell's, and I attended the Geological Society. The
Club very full. Sat between Charles Lyell and
John Moore. Much pleasant talk chiefly about
geology. The evening meeting very good; an
excellent paper by Edward Forbes, containing the
results of his long and most elaborate examination
(during the whole of last winter) of the tertiary
formations in the Isle of Wight ; very clear and
satisfactory. He showed conclusively the existence
there of several members of the tertiary series (upper

---

* She had been his Wife's governess, and was a friend whom she much
valued.

1853. Eocene and middle Eocene) that had before been
entirely overlooked, and not known to exist in
Britain; gave a methodical tabular arrangement,
and pointed out the corresponding foreign types in
France and Belgium.

_____

May 6th.

Went with Mary to see the Exhibition of the
Royal Academy. There are some interesting
pictures, but on the whole it appears to me not so
good as many I have seen. Landseer has two fine
but not pleasing pictures; one " Night," a battle by
moonlight between two stags ; the other " Morning,"
the sequel to it, in which the combatants are both
lying dead or exhausted, and a fox and a hawk
approaching to prey upon them. This latter pleases
me best; the tinge of the early morning light on the
hill-tops, the clearing mist, and the gleam on the lake,
are very beautiful, and the fox's fur is wonderfully
painted. Millais's " Order of Release " is an interesting picture, painted with infinite care, but with
less than usual of the affected peculiarities of the
astist's manner, and the accessories not offensively
prominent. The wounded Highland prisoner,
whose " order of release " is brought by his wife.
His faithful dog who is leaping upon him, and
licking his hands ; and the old weather-beaten
guard examining the order—are all excellent ; but
the wife seems wanting in expression, and the child in
her arms, though very natural, is needlessly ugly. The
same artist's " Proscribed Royalist " has more of his
mannerism ; the female figure is pleasing, but the

accessories, the ferns, moss, dead leaves, &c., are 1853.
extravagantly prominent, and finished with a super-
fluity of botanical detail that is excessively
misplaced.

One of the most interesting pictures here is by J.
Leslie, a new artist, who I am told is an officer in
the Guards ; the subject an old woman teaching to
some children the history of the Crucifixion—is
striking and unhackneyed, and the expression admirable. Stanfield has a grand picture ; the Victory
entering the Bay of Gibraltar after the battle of
Trafalgar ; and there are some beautiful landscapes
by Lee.

At the Athenæum met Sir Charles Lemon, whom
I had not seen since his illness ; he spoke cheerfully
and reported himself to be quite well, but I thought
he seemed weak. He talked of Mr. Duffey's row in
the House of Commons.

I went with the Lyell's to an evening party at the
American Minister's (Mr. Ingersell).

May 7th.

Visited Minnie and the Richard Napiers, and
had pleasant chat with them. Richard Napier gives
but a qualified approbation to the Budget ; he dislikes the continuance of the income tax, and thinks
the legacy duty altogether a bad tax in principle.
Spent much of the day at the Athenæum. Herman
Merivale talked of the mesmeric *table lifting*, which
is a great subject of curiosity just now ; he had not
seen an instance himself, and did not express any

1853. opinion as to its truth, but said that a friend of his, a member of the Athenæum, professed to have the power in a high degree. Edward said that D'Israeli's conduct last night, in voting with a fraction of the Irish Brigade against the Government, would damage the Derby party more than anything that had yet happened. Went in the evening to Lord Ross's Royal Society party, and met many acquaintances; Charles Darwin, Joseph Hooker, John Moore, James Heywood, Monckton Milnes, and others. Darwin very seldom goes out in the evening, and I had not seen him for a long time. Talked with him and Hooker about distant travels, about countries still remaining to be explored, and about Harvey's enterprising plan of going to Australia and New Zealand, to study their Algæ.

May 8th, Sunday.

Went to Church. Afterwards to the Zoological Gardens, with the Lyells, Susan and Miss Moore. The variety of marine animals in tanks of salt water, are most curious and interesting; it is a great treat to see them at one's ease, all moving about and performing their natural actions. This is all new since I was last here. There are numerous kinds of Sea Anemone, different from those one most commonly sees, some of them very brilliant, and beautiful in colour. An Ophiura, one of the long slender-armed starfishes, beautifully variegated, and most curious with its innumerable little spines, and the writhing snake-like motion of its slender arms. A curious Doris, a kind of sea-slug; Balani putting out their wonderful delicate feather-like feelers 1853. and moving them with great quickness in the water; a beautiful Echinus; some little fishes in a very lively state, and various other things. We saw also the young female Giraffe, very lately born, a graceful little creature; it is the first *female* born in this country. The Elands, from Natal, are very handsome animals; the body heavy and cow-like, but the head and eyes beautiful. In the Reptile house, we admired the beautiful motions of a very long yellow Snake, from Ceylon; and the large green Lizard from the South of Europe.

May 9th.

Drove out to Clapham Park, to dine with my Uncle and Aunt, and spent a pleasant evening with them. Besides Caroline and Norah the Arrans were there, with their little girl* of 4 years old, an uncommonly lively, clever, amusing little creature. Uncle William in unusually good spirits and humour, and very talkative. We talked about Napoleon's campaigns. He thought the greatest and most wonderful of all Napoleon's military achievements, were the series of movements and battles near Ratisbon, in the beginning of the campaign of 1809, when putting himself at the head of a retreating and discouraged army; he changed at once from the defensive to the offensive, turned and out-manœuvred the enemy, defeated the Archduke Charles, in a succession of engagements, and forced him to retreat across the Danube. Sir William thinks this the

* Afterwards Lady Ruthven.

1853. most brilliant example of military genius recorded in
history.

He thinks that Napoleon's genius was shown more
in the general management of a campaign, in com-
binations and series of movements, than in single
battles; that there was no *one* battle of his that
could be named as pre-eminently skilful, like
Rosbach and Leuthen, in the wars of Frederic,
Salamanca in those of Wellington,—The manœuvre
of the latter at Salamanca was essentially the same
as that of Frederic at Rosbach.

The Duke committed great military errors at
Fuentes de Onoro and at Toulouse.

Uncle William highly admires Marlborough's
genius, and considers the charge of his prolonging
the war for his own interests entirely unfounded;
that if he had not been hampered and thwarted by
the Dutch, he would have ended it long before
by marching upon Paris. But his ravage of Bavaria
was a deep stain on his character, and quite as bad
as the laying waste of the Palatinate by the French.

Lord Arran said that Rosas used to keep in his
house a collection of the *skulls* of people he had put
to death, carefully prepared, cleaned and polished,
and neatly labelled; and that his daughter Manue-
lita had a pet collection of her own of human *ears*,
preserved and arranged under glass.

————

May 10th.

Went with Fanny, Mary, and Susan, to a private
view of Mr. Pulsky's collection of antiquities, which

are soon to be exhibited to the public. A very 1853. curious and interesting collection, and several of the Greek bronzes very beautiful, in a very high style of art.

Mildenhall.
June 5th.

The principal thing I read was Larpent's Journal, a book both entertaining and instructive in a high degree.

Mr. Larpent was Judge Advocate General with the Duke of Wellington's army in the latter part of the Peninsula war, and his Journal (written without the least pretension, and merely for the information of his family at home), possesses a very peculiar and uncommon kind of interest, as being a near view of war by an inteligent *non-military* observer.

As Lady Bunbury says very happily, he is the *valet de chambre* of War : he shows us the *unheroic* side of it and gives us a more familiar insight into its realities than military narratives can do. The effect is certainly not favourable to War.

Mr. Larpent, moreover, though he seems scarcely to have thoroughly appreciated the great Duke, gives us many interesting familiar traits of his character and habits. Sir William Napier, when I dined with him at Clapham, commended this book as a very honest one and very trustworthy in the main, though of course not free from errors.

Besides this, I began to read the "Memoirs of Charles James Fox," but had not time to get

1853. through the volume. I was much entertained by what I did read.

We returned home to Mildenhall on the 30th, the ninth anniversary of our happy wedding; and I have since been much occupied in arranging our newly-bound books, which are numerous; in studying the Flora Antarctica and companion to the Botanical Magazine, which are new acquisitions to my botanical library,—the Flora Antarctica being a wedding gift from my darling; and in examining and arranging a large parcel of dried plants which Harvey has sent me in return for a South Brasilian set.

I hope to accomplish a good deal this summer; but—*ve dremo.*

June 6th.

Before 4 o'clock the Dean of Hereford* arrived, and after some talk we went with him to the girls' school, where he examined the children. He is a remarkably cheerful and pleasant man.

June 14th.

I have regularly and carefully studied the Flora Antarctica, which is a perfect mine of important and curious observations. I have been tolerably regular in reading Göppert, which is doubly useful to me, for the acquisition of German, and for the palæobotanical information. Being not quite satisfied with the beginning of my manuscript on fossil Ferns, I have begun to re-write it. These have been the occupations of my mornings. In the after-

* Dr. Dawes.

noons I have studied little, the fine weather 1853.
tempting me out, the company of the children,*
occasional visitors, and the newspapers, which are
just now more than usually interesting from the
Indian discussions, and the Turkish crisis,—together
with no small amount of *dawdling*, have run away
with my time. In the evenings in reading aloud;
we have only got through the first chapter of
Gibbon.

June 16th.

To Hardwick; met Edward and Mr. and Mrs.
Thomas Mills, of Stutton, near Ipswich.

June 17th.

Donaldson's lecture on the Etruscan tomb at
Hardwick.

June 18th.

Mrs. Jameson arrived, and after dining at
Hardwick, we took her to Barton, where we
slept.

June 20th.

Shewed Mrs. Jameson the arboretum, and after
luncheon returned by way of Bury to Mildenhall.

June 22nd.

Mrs. Jameson left us. She is a remarkably
agreeable woman, highly cultivated and accomplished, with much originality and vigour of mind,
great liveliness and a considerable share of humour.

* His brother Hanmer's children were staying with him.

1853. Her conversation is as excellent as her books. Her acquaintance with the fine arts appears to me very extensive, but she disclaims any technical knowledge of pictures; she says that she cannot pretend to know the work of any particular painter by the colour, the touch, or other points on which connoisseurs affect to dwell; but she thinks that the great masters may be known by the *feeling*, the expression of their moral or intellectual *character* peculiar to each, and always more or less visible in their works. She says, very truly, that there is no reason why one should not understand and judge of paintings, as one does of poetry, without descending into technical minutiæ. She is full of interesting anecdotes and illustrative information, relating to artists and pictures. The condition of women in society, and the various moral and social questions relating to them, are subjects on which Mrs. Jameson feels strongly, and is fond of talking. I heartily agree with her on many points of this delicate matter.

LETTERS.

July 18th, 1853.

My Dear Mary,

It is delightful to hear of your being so well and happy in America, as your letter shews you to be, but I hope you will take care of yourselves, for the newspapers report the heat at New York to be terrific. Pray beware of too much exposure to

the sun. I wish you could send us over some of the 1853.
heat which it seems is to spare there ; for we are
shivering in one of the most ungenial summers
I remember. Nothing but rain and stormy gales,
almost without intermission, and from every point of
the compass, so that we have not even the comfort
of hoping that a change of wind will bring an improvement. It is certainly the worst summer we
have had since 1845. It feels more like the end
of October than the middle of July. Our roses,
which began in great beauty, are sadly knocked
about, and what is of more consequence, the prospect is very unpromising for the harvest.

This is my darling Wife's birthday.

I pass my time very pleasantly and comfortably
in my museum, &c. I am hard at work (whenever
the weather is not too dark), making drawings of the
genera of fossil Ferns for my book. I am going on
pretty steadily, though by little at a time, with
German.

I have been reading with great pleasure Mrs.
Jameson's "Sketches of Canada," now republished
in a cheap form. It is very prettily written, like all
her works, and while very lively and entertaining, it
is far from flimsy, but contains much matter for
thought. She describes beautifully. There is a
description of the falls of Niagara as seen in the
depth of winter, that is most striking ; and her
account of the navigation in a canoe among the
innumerable islands of Lake Huron gives one a
longing to go there. I knew nothing about those
upper Lakes. I daresay that some of these days

1853. you will be off to Michilimackinox and the South Sainte Marie. I like Mrs. Jameson very much,— not only her style of writing and her conversation, but her sentiments and opinions on most points. She was very agreeable, though she did not seem in good health or spirits when she was here, and I was very sorry she was obliged to leave us so soon.

We have been very anxious and unhappy about Sir Charles Napier, who has been very dangerously ill; so much so, that at one time his death was expected hourly; but he rallied in a manner that astonished all the medical men, and though not completely out of danger, he is now so much better that there are great hopes of his perfect recovery.

Lady Bunbury, herself very far from well, went up to town last week to see him, and I am afraid she will be much the worse for the exertion and anxiety. This alarming illness of Sir Charles cut short the William Napiers' visit to us, and has entirely prevented Minnie's, which we were reckoning upon.

We are still in suspense about the great question of peace or war, but on the whole it seems most probable that there will be peace, not, I fear, without some truckling to Russia and some increase of her reputation and influence. War,—such a war as this assuredly would be if it once began,—is a fearful thing to contemplate; and yet I am almost inclined to wish for it, for I believe that the struggle with Russia *must* come, and that it is not likely ever to come under circumstances more favourable to us. Russia, as Mr. Grote hints, is now to the rest of

Europe what Macedonia was to Greece in the days 1853.
of Demosthenes. Besides a general war is the only
chance of relief for the oppressed nations under the
Austrian yoke, and it is evident that Austria sees
this by her extreme anxiety to mediate.

I wonder whether the New York Exhibition has
opened yet ? And if it has, I suppose you are both
extremely busy. I hope however in due time to
hear of your splendours.

Fanny sends much love, and with mine to Charles,
Believe me ever, yours very affectionately,
C. J. F. BUNBURY.

———

From his Father.

Barton.
August 21st, 1853.

My Dear Charles,

I cannot send the accompanying letter
without telling you how deeply I have been gratified
by your warmly affectionate letter of Friday. My
cordial thanks to you for it, my dear Son. Our last
account from Oaklands is of Friday evening:—Sir
Charles was still alive. His wonderfully strong
constitution struggles hard against death. Emily
went yesterday to Clapham, to be more within reach
of intelligence; but I think she will come back in a
a day or two.

I am indeed grieved at the sudden loss of my old
friend Adam.

Ever affectionately yours,
H. E. B.

Mildenhall,
August 29th, 1853.

My Dear Lyell,

We are enjoying Susan's visit exceedingly ; she is a delightful companion, and she is going on most capitally with Fanny's picture. I think it very like and a particularly agreeable likeness.

Our weather is rather Fuegian, but nevertheless we have had some very pleasant drives and *asines- trian* excursions. My Fanny is, I flatter myself, certainly gaining strength, for she is able to walk much more than she could do some time ago, though she is not in all respects quite well. The season has been a very trying one. Poor Sir Charles Napier still lingers, paying a cruel penalty for the strength of his constitution ; such a long death struggle is dreadful both for himself and those about him. He is a sad loss ; such a warm-hearted, kind, affectionate, excellent man, that to his family and friends he can never be replaced ; to say nothing of the great and heroic part of his character. How the men distinguished in the last war are passing away !—Sir Frederick Adam, Sir George Cockburn, Lord Saltoun, all dead so nearly at the same time, and Sir Charles dying. The Turkish question seems likely to be patched up for the present, but I think war must come soon,— and the sooner the better.

I have lately read over again your paper on "Craters of Denudation," and I am very much struck with it. You make out a very strong case. With what interest you will examine the crater of

Palina! I daresay you will throw a great deal of 1853.
new light on it after seeing it with your own eyes,
and I expect that your Canarian expedition will be
as profitable to the geological world as delightful to
yourself.

With much love from both of us to Mary, I
am ever,

Yours affectionately,

C. J. F. BUNBURY.

Mildenhall,
September 6th, 1853

My Dear Lyell,

In lately looking over some of the chapters
of the third book of your Principles, I observe
that you touch upon the question of the origin of
Wheat.

Have you chanced to hear of the strange discovery
lately made by a Frenchman, of the name of
Fabre, who affirms that he has ascertained that a
grass called Ægilops ovata is transformed by cultivation into wheat? His experiments are recorded
in great detail in a paper which is translated in a
late number of the Journal of the Agricultural
Society. Of course all depends upon the accuracy
and trustworthiness of the observer; but his experiments appear, on the face of them, to have been
careful and judicious, and Dunal, an eminent
French botanist, supports his conclusions. I asked
Hooker about it when he was here; he said he had
not himself had the opportunity of investigating the

1853. subject, but that Bentham and Lindley both believe the fact.   If true, it is very curious, for the Ægilops in its natural state is certainly not very like Wheat, though it belongs to the same main division, that of spiked grasses with a jointed *rachis*. It is an annual grass, common in all the Mediterranean countries, and most particularly in *Sicily*; you remember the connexion of the myths of Ceres with that island ?   By the way I should mention, that there is a variety, or what some botanists considered a species, occasionally, though rarely, found wild, which was called Ægilops *triticoides*, from its striking resemblance to Wheat.   After all, the fact, supposing it well proved, though very curious, does not go far to help the Lamarckians; it merely proves that the range of variation in some species, is greater than what naturalists have generally been aware of; it does not at all go to prove an *unlimited* capacity of *progressive* variation.   That Wheat was not, in the fullest sense, a natural production, might perhaps have been suspected from its never being known in the most favourable soils or climates, to establish itself to any extent as a naturalized plant, or to hold its ground long after careful cultivation has ceased. No, to be sure, my inference is not quite sound; this might only prove the delicacy of its constitution. We have a most remarkable instance in this country of a plant naturalized within the last few years, and already spreading to such an extent, as to become a very serious nuisance.   It is a little American water plant, called Anacharis alsinastrum, (and several other names besides).   Till within the last five or six years it was unknown in England, 1853. and now several of the fen rivers are choked with it to such a degree that the navigation is seriously impeded.

This year for the first time it has appeared in our parish. Its rapidity of increase is truly wonderful. I first saw it on the 13th of July, and by the 26th of August it had spread so that where there were on the former day only straggling plants, it now formed dense broad masses. A water cut, where I am certain that last year there could have been little if any of it, is now almost completely choked. Very many weeds have been introduced with cultivation from Europe into other continents but this is the first instance I know of an exotic, *not* a weed of cultivated land, establishing itself so rapidly to such an extent in this country.

I quite agree with you that the conduct of our Government towards Hungary was wrong from the beginning, but I do not see how it is to be righted just now; at any rate I am sure that public opinion would not have supported the Government in declaring an *unprovoked* war against Austria, which would indeed have been contrary to all the laws of nations. A general system of revolutionary propagandism (even supposing it justifiable) cannot be well or safely undertaken except by a country that is itself in a state of revolution, as France in '92 and the following years. It might have been better on the whole if Austria had joined Russia, but I do not think our ministers can be blamed for not *driving* her into that alliance; though I certainly think they

1853. have made too many sacrifices for the sake of pleasing her.

I am afraid the movement in Spain will lead to no permanent good; nothing short of sending the whole Bourbon Family on their travels will secure that country against their perfidy and wickedness.

With much love to Mary, believe me ever,

Yours affectionately,

C. J. F. BUNBURY.

Mildenhall.

September 13th, 1853.

My Dear Katharine,

Your letters are always very agreeable, and I am truly glad that you are able to give such comfortable accounts of yourself and Harry and your dear little ones. My last letter to you was written at Ventnor, the end of March. Susan left us yesterday morning, after most kindly paying us a visit of more than three weeks; and a delightful visit it was to me. I do not know a pleasanter companion than she is, so full of cleverness and knowledge and good conversation, so cheerful, so kind and warm-hearted. Susan has most kindly painted for me a picture of my dear Wife, which now occupies the place of honour in my study, and is a very great ornament to it. *I* think it very like; of course there will always be a diversity of opinion on such a point, but it is assuredly a very pretty picture, and very well painted. I was very sorry to part with Susan, but we are now expecting the pleasure of a visit from your dear Father and

Mother and Leonora, who are coming to us next 1853. Monday on their way to Manchester. We have lately been much grieved by the death of Sir Charles Napier : he was ill nearly two months. He is a sad loss : he was even more a good than a great man. His country will feel the loss of his heroic qualities, his commanding talents and high reputation; but those only who knew him personally, know how full he was of kindness and warmth of heart and genial sympathy; how unselfish and generous and affectionate; how warm and true a friend. Fanny and I loved him very much. You will see in the newspapers the account of his funeral. It was very interesting to see so much *spontaneous* feeling on the part of the people, and especially of the soldiers : 3000 soldiers volunteering to attend the procession. It shewed that his merits were really appreciated, though not by the people in authority; and it must have been a gratification, though a mournful one to his family, to see such a voluntary and unsolicited tribute of affection and respect.

How the men who were conspicuous in the last war are passing away! It is but just a year since the Duke died, and now within the last month or two, Sir Charles Napier, Sir Frederick Adam, Sir George Cockburn, and Lord Saltoun, are gone.

I do not know that I have very much to tell you in the botanical way. But in the first place I must thank you very heartily for your kindness in sending me specimens of the wood and leaves of the Banyan tree, which I value very much. I have examined the wood microscopically. I hope

1853. you will have had good botanical sport at Simla, which I should think must be an interesting place in that way, though I suppose not equal to Darjeeling. I have been a good deal occupied lately with microscopic examinations of the internal structure of plants, having got a new and very good microscope, which was made for me this spring, under the direction of Joseph Hooker. I hope to show it you some of these days. In particular, I have been busy examining Equisetums, as I am engaged with the history of that family; and I find much that is interesting in their structure. The *epidermis* is a beautiful microscopic object. It is remarkable how much that is interesting one finds in common plants when one comes to examine them closely and carefully. A fine work by Dr. Torrey, on the botany of New York, which Mary has lent me, has been of material service to me.

Besides this I have been occupying myself a good deal in drawing fossil plants, and I have been reading a good deal of geology and a little German.

I saw in the Athenæum some time ago a curious account of the valuable properties that the Asclepias gigantea has been discovered to possess; I think you alluded to them in your former mention of the plant in one of your letters. It seems likely to be a very important plant, as affording both a kind of Gutta-percha and a valuable fibre.

The last number of the "Geological Society Journal" contains a geological description by Dr. Fleming, of the Salt Range of the Punjaub, with some curious observations.

The season has been very unfavourable; the most 1853. wintry summer almost that I ever remember; nothing but rain, wind, clouds and gloom; generally very cold, and when warmer, damp, muggy, unhealthy weather. Now at last we have some really beautiful autumn weather. The cholera has made its appearance at Newcastle, and the unfavourable season has doubtless well prepared the way for it, but as it has come so late in the year, I hope it may not spread very much.

Cecilia has a little girl, and is reported to be doing well. I trust your little men are thriving; they must be amusing little fellows.

We shall probably go abroad this winter, but whither is not quite certain. We sometimes talk of going to Madeira and the Canaries with the Lyells: it would be a very pleasant expedition and a delightful climate to pass the winter in ; *ma vedremo.*

Ever your very affectionate Brother,
C. J. F. BUNBURY.

Mildenhall,
September 27th, 1853.

My Dear Lyell,

I shall be extremely glad to see your new edition of the Manual, and have no doubt that I shall learn much from it, especially as to the Tertiary formations, with which I am certainly less conversant than with the rest. The classification of those formations must be very difficult, more especially because they seem (with the exception of the Nummulitic formation) to be more fragmentary and disconnected, to have more of the character of

1853. local deposits, than the older rocks in general. I
have no doubt you have judged rightly in fixing your
line of division between Miocene and Eocene,
where you say you have: for the unfortunate
Miocene seemed in most *immiment* danger of being
swallowed up, "blandly absorbed," as Sydney Smith
would say, or partitioned between Eocene and
Pliocene, as Russia and Austria would like to
partition Turkey. But after all, is not the line of
demarcation (as depending entirely on fossils)
necessarily rather arbitrary ? Is it not likely
(especially on the *uniformitarian* theory), that the
progress of discovery will tend to remove all absolute
distinctions between one of these systems and
another ? Is it not particularly necessary in judging
from zoological considerations, to take into account
the differences occassioned by *place*, as well as by
time ; And in particular, is there more difference
between the Touraine Faluns and the Suffolk Crag,
than between the recent Fauna of the Mediterranean
and of the Red Sea ? I know how thoroughly you
have studied and considered these points, and I do
not at all mean to criticise your arrangement, but
simply to suggest the doubts that occur to me.

It seems very doubtful, as yet, how far the (very
local) fossil Floras of the tertiary period can lend
any aid towards the geological classification of the
deposits. It is not easy to be certain of a plant (a
mono or dicotyledonous plant) from detached leaves,
or even from single fruits if these retain no internal
structure. Still I think it would be well worth while
in your Manual, just to notice the most remark-

able tertiary deposits of fossil plants, such as those
of Œningen, of Parsschlug in Styria, and of various
localities in Croatia. By the way, Unger refers all
the Œningen plants to the Miocene, Adolphe Brongniart to the Pliocene : I suppose the latter is right.

I am reading Lepsius with great interest. I am
particularly diverted with the custom existing in
some parts of Nubia : that when the king ceases to
be popular, his subjects and even his own family
politely and blandly persuade him to be hanged,—
to which he seems to submit with the utmost
equanimity. I wonder what King Bomba would say
to this! It is certainly an improvement in point of
politeness, upon the plan pursued towards Charles
I. and Louis XVI.

I am very anxious to see what will be the course
of our Government, now that Russia and Austria
have thrown off the mask, and Turkey stands so
manfully on its defence. If we flinch and truckle
now, allowing ourselves to be insulted, and laughed
at by the Czar, we shall be degraded indeed, and
sink (deservedly) into a most insignificant position
as regards the affairs of Europe.

But I am not hopeful: I have no confidence
whatever in Lord Aberdeen, and I fear that his
cringing and truckling propensities will be but too
much encouraged by the Manchester School, and by
the shopkeeper spirit so prevalent in England. The
Americans seem well-disposed to take up the part
which we have dropped, of protectors of con-
stitutional freedom in Europe, and I daresay they
will fill it very efficiently. I expect that before

1853. another generation has passed away, the influence of America upon European politics will be very considerable. I was delighted with their spirited conduct at Smyrna.

I was heartily sorry to part with Mr. and Mrs. Horner and Leonora so soon, having enjoyed their short visit very much.

I long to see you and dear Mary, and you ought certainly to come hither, if it were only to see Susan's charming picture of Fanny, before you go to the Canaries. We talk now seriously of going with you as far as Madeira at least; that island would be a better and safer quarter for one so delicate as Fanny, than the less civilized Canaries; and it is a most lovely island, and though certainly less remarkable in its botany and geology than Teneriffe, must have much that is interesting. I should like of all things to see Teneriffe, but I do not know whether we shall be *up* to that.

Fanny is tolerably well, I think, and goes much about in her pony chair, which is a great comfort to us. She sends much love to both of you, in which I heartily join.

Ever affectionately yours,

C. J. F. BUNBURY.

JOURNAL.

September 28th.

Began, in compliance with Arthur Hervey's wish, to write a discourse on the study of natural history for the Bury institute.

October 1st. 1853.

Finished my discourse on the study of natural history which has occupied me a good deal for these last few days, and given me some trouble.

October 4th.

Opening of the Bury Athenæum and Institute of Archæology, and Natural History. Arthur Hervey the President, made a beautiful address, admirable in feeling and expression. My discourse was very well received, and I was glad to find that it gave great satisfaction to Hervey as well as to my Wife.

Mr. Sidney gave us a clever and amusing address. Sir John Walsham, Donaldson and Dr. Hake also spoke. Altogether it was a satisfactory meeting, though the extreme wetness of the evening kept many away.

October 5th.

Dined at Lord Bristol's. He is an extraordinary man of his age (84) ; active and vigorous in mind and body, with all his faculties in complete preservation. He told me two curious instances of predictions unexpectedly verified. The first when Louis Napoleon was an exile in England, before the revolution of February, 1848, he was present with many other fashionable people, at the marriage of Miss Damer to Lord Ebrington.*

After the marriage, the fashionable guests proceeded to sign the register as witnesses. Mr. Seymour (Lord Bristol's son-in-law) observed Louis Napoleon rather standing aloof, and asked him why

* March, 1847.

1853.   he did not also sign with the rest. " What use would
it be ? my signature cannot matter here." " Oh,"
said Mr. Seymour, "*you will be Emperor one of these
days*, and then your signature will make the clerk's
fortune."   Certainly at that time it did appear
exceedingly unlikely that Louis Napoleon should
ever be Emperor.

The other story was this; when Louis XVIII. was
an exile in this country, he went one day to see a
man (a foreigner) who pretended to tell fortunes by
the cards.   The man said to him :— " Vous serez
roi, mais vous ne serez jamais sacré."   And it did
so turn out, that Louis XVIII. during the whole of
his reign never was " sacré," for he was too unwilling to go to Rheims, and had a scruple against
having the ceremony performed at Paris.

October 7th.

Finished reading Lepsius' letters, which make up
about two-thirds of the volume translated by
Leonora and Joanna.

LETTER

Mildenhall,
October 16th, 1853.

My Dear Mr. Horner,

We have been much grieved by the death
of poor Captain Henry Napier; not that one ought
perhaps to regret it on his own account, for to one
whose life, for many years past had been one long
and painful illness, with little or no hope of amend-
ment, death must surely have been welcome.   But I

had a great regard for him.   He was a truly kind, 1853.
warm-hearted, generous, noble-minded man, one of
those whose loss must always be regretted.   It is a
comfort that he lived to see his daughter so happily
married.   Poor Augusta* will feel his death sadly,
but I feel almost more for Lady Bunbury. who has
lost within so short a time two brothers whom she
tenderly loved.

I have just finished Lepsius, which I have read
with great satisfaction.   The letters are very interesting, and I think the translation remarkably well
written.   Your paper on the supposed sinking of
the level of the Nile is very curious and important,
and I think you have pretty clearly made out the
impossibility (or at any rate the extreme im-
probability) of such a change being produced within
the time assigned by the mere ordinary and
gradual operation of the river.   The data you have
brought together in that paper are valuable and such
as, I should think, are not easily to be found elsewhere.   Lepsius's answer does not appear to me
satisfactory, in fact I do not clearly make out what
he means.

The look of affairs in the east is now very warlike
indeed.   I hope the Sultan will not allow himself
to be shackled or blinded by any more conferences,
as I take it there is nothing will suit the Czar's
purposes better than further delay.   Mr. Cobden
must really reckon very much on the gullibility of
his hearers, when he tells them that the acquisition
of Turkey will rather weaken than strengthen

* His Daughter, who married Frederick Williams Freeman.

1853. Russia.   He is much nearer the truth in saying that our encroachments in India have been as unjustifiable as those of Russia in Europe and Western Asia.   But all civilized nations seem much alike in their dealings with " barbarians."

With much love from Fanny and me to Mrs. Horner and Leonora as well as to yourself,

Believe me ever,

Your affectionate Son-in-law,

C. J. F. BUNBURY.

JOURNAL.

October 19th.

Began reading Arnold's Lectures on Modern History.

November 3rd.

Took a long drive through Mildenhall Fen with Mr. Cooper, quite to the extremity of my Father's estate.   Nothing can be more drearily and monotonously-ugly, or more uninteresting (unless to the eye of a farmer) than this reclaimed Fen.   The thick, spongy soil (though thoroughly drained) has so little solidity, that houses, unless of wood, or of lath and plaster, cannot be safely built on it, without first preparing a foundation by means of piles.   In one place, where part of a cottage had given way, the man told us that the depth of the black peat soil was 11 feet.

The reclaimed fen land is found to answer better for grass than for corn.   It produces indeed

apparently good crops of Wheat, the plants are 1853.
abundant; but they are little but stalk and leaf;
the ears do not fill, *i. e.* the seeds are in great part
abortive; at least such has been the case for some
years past. It appears, however, from some
experiments which have lately been tried under Mr.
Cooper's direction, that the application of *lime* as
a manure is of great benefit. The best grasses to
cultivate in the Fen land, I am told, are Timothy
grass (Phleum pratense), and Italian Rye grass
(Lolium Multiflorum). Turnips do not succeed,
but Cabbages and Beet do very well.

As the proper treatment of the Fen land requires
considerable capital, it is best laid out in large
farms; and it is found advantageous to allot these
farms to tenants who have large " upland " farms
(on the dry soil that is to say), so that each of the
principal tenants has a large Fen farm, as well as a
large farm in the other division of the Parish.

LETTER.

Mildenhall.
November 14th.

My Dear Mrs. Horner,

We have have had a very delightful though
short visit from the dear Lyells, who seem in
excellent health and spirits, and were incomparably
agreeable. I do hope and trust they will not give up
their Canary expedition, though I cannot help
having some fears, especially since Joanna seems
uncertain about accompanying them. It would be

1853. *so very* much pleasanter our all going together;
and I have been thinking of it so long, I can hardly
bear to give it up.

Pray give my love and congratulations to dear
Leonora, to whom I most heartily wish all the
happiness that a wise and well assorted marriage
can give,—and that is about as much as this life
can give. I am sure she is well-deserving of every
blessing, and though marriage is always something
of a lottery, yet, entering upon it with so wise and
thoughtful a spirit, and with such feelings and
principles as hers, I think she has as fair a prospect
as can well be of happiness.

I am very desirous of becoming acquainted with
Dr. Pertz, whom I do not doubt I shall like
much. I hope Leonora does not mean to give up
her botany, as she will be in the neighbourhood of
of one of the finest botanic gardens and botanical
museums in the world, and doubtless will have the
opportunity of knowing all the scientific celebrities
of Berlin.

Give my love also to Mr. Horner, and to Susan
and Joanna, and believe me,

Your very affectionate Son-in-law,

C. J. F. BUNBURY.

JOURNAL.

November 20th.

As we are preparing very soon to leave England
to spend the winter in Madeira and Teneriffe, I will
note down here something of what presents itself

to my mind on looking back upon the year which is 1853.
now nearly past. And first I will acknowledge what
great reason I have to be thankful to Almighty
God, for the many and great blessings I continue to
enjoy. That my home is still rendered bright and
happy by the presence of my admirable wife ; that
her health is gradually improving while our mutual
affection and confidence, and unity of feeling grows
stronger in proportion as we live longer together ;
that all those who are especially near and dear are
still spared to us ; and that my own health though
not very strong, continues sufficiently good to allow
me much enjoyment of life, and the free use of my
mental faculties ; these are things for which I
cannot be sufficiently grateful.

This year indeed has been saddened by the loss
of two excellent friends—two truly warm-hearted
and excellent men—Sir Charles Napier and his
brother Henry. Sir Charles was a public loss, and
much as he had been abused and misjudged during
his life, his death was felt as a public calamity.

But those who knew him personally, knew that he
was more a good than a great man. No man was
ever a truer friend or more worthy to be loved.

His brother, with inferior talents, and the disadvantage of an unhappy and morbid temper, had
the same nobleness of mind and energy of affection.
Both were most cordial and affectionate friends to
us.

During the summer and autumn from London,
we have enjoyed very delightful visits from several of
our friends, whose society has been equally pleasant

1853. and profitable to us; though from one cause and another, and especially through the death of poor Sir Charles Napier, we have been disappointed of some that we reckoned upon.

First after our return home came the Dean of Hereford; next Mrs. Jameson; then the William Napiers, whose stay with us was cut short by the news of Sir Charles Napier's danger; next Susan Horner, who spent a considerable time here, and was exceedingly pleasant; then Mr. and Mrs. Horner and Leonora, for one week only on their way to Manchester; Mrs. Charles Young; the Charles Lyells who could spare us only three days; and lastly my Father and Lady Bunbury, Henry and Cecilia. I have omitted to mention Sally and her children who have been here more than once.

When I look back on the six months spent here, and the comparative leisure and quiet I have enjoyed, I cannot help feeling, as unfortunately I have very often had occasion to feel, that I have done but little. I have indeed been, though not positively ill, yet in a languid and feeble state of health, and have had even more than usual difficulty in bracing myself up to any exertion. I have pursued various schemes in the scientific way, and not thoroughly carried out any of them. I do not feel that I have made any material progress in the study of German, a language which to me is almost unconquerable; and though I have certainly added considerably to my knowledge of various branches of botany and geology, I have nothing very positive to show by

way of result. The most material thing I have done 1853.
has been writing the whole of the chapter on Equisetaceæ for my Genera of Fossil Plants; in doing
which I have made myself pretty extensively acquainted with the structure and peculiarities of that
curious tribe, which I before knew but superficially.
I have written several portions of the Genera of
Ferns for the same work, and intended to have gone
through and corrected the whole of that part of the
subject, and worked it up to as complete a state as I
could, but this I have not accomplished.

LETTERS.

Funchal,
December 18th, 1853

My Dear Father,

I wrote a few lines to you from Lisbon, on
Wednesday the 14th, telling you of our safe arrival
there, of the very rough weather we had met with,
and our apprehensions of a rough passage from
thence. Happily these anticipations were not ful-
filled; the next day proved fine, and we went on
smoothly, and altogether, though the wind has
not been very favourable and the "Thames" not
a very fast steamer, we had a pretty good passage
from Lisbon hither, and suffered no more from
sea-sickness. We came in sight of Madeira late
last night,—too late to come in, so we lay by during
the night, and were rather uncomfortably tossed
about; and this morning we entered the Bay of
Funchal in a drizzling rain, while the clouds hung
so heavy and so low on the hills as to give a rather

1853. gloomy aspect to the scene. But the appearance of
Funchal from the sea, with its bright white houses
so picturesquely clustering up the mountain sides, is
always very striking; and towards the middle of the
day the clouds retired to the higher mountain
regions, the sun shone out and the weather became
beautiful.—Young Chichester, the son of our Barton
Mills neighbour, came on board almost as soon as
the vessel anchored, greeted us heartily, and was
most truly kind and zealous and active in helping us.

The mode of landing is curious; as the shore is
excessively steep and the surf heavy, it is rather
a difficult operation; the boat is pushed in pretty
near to the shore, but not aground, and then as it is
rocking on the waves, a number of men rush into
the water, seize hold of it, and by great strength
and dexterity fairly haul it ashore. How they
escape capsizing it seems a marvel, but it is said
that accidents never happen. The carriages used in
Funchal, and in one of which we went lodging
hunting, are most original contrivances, unlike anything I have seen in any other country; a sort of
large covered sledge without wheels, very gaily and
smartly fitted up and drawn by a couple of bullocks!
They go sometimes at a marvellous pace, the driver
walking or running by the side of the animals with
a long stick. I should like amazingly to see the
face of Walter Wallis* contemplating such a *turn-
out*. These quaint conveyances are comparatively
new in Madeira; there were none when I was here
twenty years ago.

* His Father's coachman.

Then as we passed through the town, our eyes 1853. were caught at every moment by remarkable plants and beautiful flowers in all the gardens : the most splendid blue Ipomeas and orange Bignonias covering the trellises and hanging over the garden walls; the Poinsettia and the Datura arborea in profuse bloom in the open ground ; Bananas, Coffee bushes, and all sorts of tropical plants, except Palms, of which I see but few.—We, the Lyells and ourselves, have hired for a week a set of rooms in the house of a Portuguese merchant; very pleasant rooms, spacious, very clean and very airy, in a fine situation. The view from our sitting-room windows is indeed glorious, ranging over the steep slopes covered with gardens of Eden-like luxuriance, interspersed with bright, gay villas like those of Genoa, to the wild, dark rugged mountains, whose tops are lost in the clouds, while in another direction the sea comes in to complete the view. Madeira is even more beautiful than my recollections pictured it. And then for the climate,—this is the 18th of December, and certainly we had very few days in the whole course of last *summer* in England equally warm ; I am sure not one evening as warm as this evening has been. The air indeed is moist and muggy, reminding one of Devonshire summer weather. They say however that there has been an unusual quantity of rain lately, and that we may hope for a spell of fine weather.

December 20th.—Yesterday was quite a heavenly day, the most glorious I have seen for a very long time. I despair of giving an idea how ex-

3 M 2

1853. quisitely beautiful the country looked under its influence. I took a long walk in the morning, up to Nossa Senhora do Monte, but had not much botanical success, as I could not disentangle myself from the stone walls and paved roads; I found however a few interesting plants. After dinner we sallied forth on another excursion,—Mary Lyell and I, young Chichester and a friend of his, all on horseback, Fanny carried in a hammock slung to a pole, and carried by two men,—a common mode of travelling here. We had a very pleasant hour's ride along the mountain side, round the heads of several valleys, enjoying the beauty of the evening and the lovely scenery glowing in the light of the setting sun. In short the day passed delightfully, and we are all in love with Madeira. Our quarters are very comfortable; it is only a little inconvenient that no one in the house, except our host, understands English, and we have not yet got up our Portuguese. We board with the family, which is a little constraint, but they are extremely civil and obliging, and we fare very well. We feast on Bananas, which are abundant and excellent, as good I think as in Brazil.

I trust we may have by the next mail a good account of you and Emily, at least that you are as well as it is possible to be in an English winter.

It is very pleasant travelling with the Lyells, they are so cheerful and full of resources, so clever and so good. This day, the 20th, has been very showery but very warm and bright between the showers, like Devonshire summer weather; and we are now at

night sitting in a large airy room, without feeling at 1853.
all the want of a fire. Much snow however has
fallen on the higher peaks of the mountains.

In our ride to-day, just as one of the showers was
passing away, we saw a most gorgeous rainbow,
stretching in a curve of vast extent along the face of
the mountains, seeming as if it lay directly against
them, and producing an effect really magical.

December 21st.—Our host tells us that within the
last five years, 30,000 persons have emigrated from
this island to Demerara and the West Indies. This
is a greater emigration, in proportion to the population, than that from Ireland. Two-thirds of the
number, it is said, are already dead, but the remainder are growing rich. The peasantry here
seem very poor, but a good, mild, obliging, cheerful
set of people, very industrious,—great beggars all
the same. The failure of the vine-crop has of
course told heavily upon all classes, but most of all
upon the small proprietors. But even before the
grape-disease came, I am told that the incomes
of the principal merchants had been considerably
effected by the change of fashion in England with
respect to wines, the demand for Madeira being so
greatly diminished.

December 23rd. This is again a lovely morning.
It seems so wonderful, after what one has been used
to in England, to open one's eyes morning after
morning to a blue sky and a bright sunshine. Yesterday and the day before, indeed were very showery
but warm and bright between the showers,—very

1853. good English *summer* weather.   There is no going
however at present to the high mountains, whose
tops are covered with snow and generally buried in
clouds.   Next month they say will be more favourable for expeditions both to the mountains and
to the north side of the island.

I wish you with all my heart a happy new year,
and am ever

Your very affectionate Son,
C. J. F. BUNBURY.

Funchal,
December 20th, 1853.

My Dear Emily,

It would require a much more eloquent
pen than mine to do justice to the fairy gardens of
Madeira.   Imagine the plants of all the tropical and
temperate parts of the globe flourishing together in
the richest luxuriance ; the choicest plants of our
hot-houses luxuriating as if in their native country ;
the Poinsettia in splendid perfection, covered with
its gorgeous crimson floral leaves, the Datura,
loaded with its great white bells ; the Bignonia
venusta covering the garden walls and trellises with
glorious festoons of its brilliant orange-coloured
flowers, as luxuriant and beautiful as ever I saw it in
Brazil ; Magnolias loaded with ripe fruit ; Australian
Acacias and Eucalyptuses ; Sugar-canes, Bananas
as plentiful as cabbages in English gardens ; Coffee
bushes, and many more than I can enumerate—to
say nothing of such *vulgar* things as prickly pears

and huge orange trees. The richness, the Eden-
like beauty of the whole effect is beyond what words
can express. And these gardens which cover the
steep slopes ascending from the sea are backed by
the dark, wild mountains of strange rugged volcanic
aspect, some of whose tops are now covered with
fresh-fallen snow. In our rides we can hardly go
fifty yards together without bursting out into fresh
exclamations of " beautiful ! beautiful ! Oh how
beautiful !"

On reading over what I have written, I am afraid
you will think me rhapsodical, but really I do not
exaggerate. Madeira might well be called as the
Chinese call their country, " the Flowery Land."
The soil and climate are most wonderfully adapted
to the productions of all parts of the world. Observe
that I have been speaking entirely of cultivated
plants ; of the native vegetation of the island, I
have been able to see very little. And here is the
disadvantage of the place to a botanist ; the cultivation extends so far, that it is very hard work to
get clear of the endless stone walls and paved roads,
and to have a glimpse of unsophisticated nature.
Everybody here rides, and even I, though by no
means partial to that mode of progression, have
been obliged to give in to the fashion, for the pave-
ment of the roads is cruelly harrassing to one's feet,
and it is a waste of strength to walk for miles and
miles between stone walls. The roads are steeper
than an inhabitant of East Anglia could conceive in
his wildest dreams ; Fanny is carried in a hammock, a common mode of conveyance here,

1853. and both a picturesque and a luxurious one. There is one coniferous tree in the gardens here which strikes me very much by its grace and beauty ; it is a weeping Cypress which the people call the Cedar *(Cedro)* but it is, I believe, a true Cypress, a Cypress of Goa, Cupressus glauca if I am not mistaken. I will try to get cones of it, but I am afraid it would be tender in England. It has something of the look of the Deodar, but much more decidedly weeping.

(December 21st). We have had a pleasant ride to-day—Mary Lyell, Fanny, young Chichester and I—to a place called Palheiro, where there are large woods (planted) of Pinasters, at some height above the level of Funchal. We got to some wild ground and had a little botanizing, and I was delighted at finding abundance of the beautiful Hare's Foot Fern, Davallia Canariensis, growing on low stone dykes, and at the roots of Chesnut trees. In one place a great patch of ground was covered with a Cape Pelargonium (the old Rose scented Geranium of gardens) completely naturalized and to all appearance wild ; and close by was another large patch of a beautiful Oxalis, I believe also a Cape plant and naturalized.

I hope you will not be quite out of patience at the quantity of botany I have given you in this letter, but I know you have, like myself a liking for such things, and I think you would enjoy Madeira if you could be brought hither without undergoing the sea voyage. Some of our fellow-passengers were pleasant, in particular, Mr. and Mrs. Aubrey Spring

Rice, whom we have seen several times since we 1853.
landed.

Before you receive this letter, the new year will
have begun. That it may bring health and happiness to you and my dear Father, is my earnest
prayer. I hope you are by this time in Wales,
which I know always agrees with you much better
than Barton. Pray give my love to Cecilia and
Henry.

Believe me ever,

Your very affectionate Step-son,

C. J. F. BUNBURY.

JOURNAL.

At Camara de Lobos, on the shore, a thick bed
of very characteristic volcanic cinders, very slaggy,
lying on a bed of the hard compact basalt. Crystals
of augite in the cinders. The basaltic lava I have
now observed in several places : in the sea-cliffs by
the Pontinha, and in the sides of the ravines both on
the way to Palheiro, on the East, and to Camara de
Lobos on the West. It seems very uniform in
character : forms thick and extensive beds, with a
rudely prismatic structure, of a deep blue-grey
colour, very hard, heavy, and sonorous, generally
very compact, but sometimes with a few small
vesicular cavities.

Funchal,

December 24th.

We went to Praya Formosa, by the new-coast
road, the same by which we had returned from

1853. Camara de Lobos. Pass along the southern base of Pico da Cruz. Fine view of the grand precipitous headland, Cape Girao before us: a precipice 1,600 feet high.

Descend to the beach at Praya Formosa, by a steep path, along the sides of which I find a species of Sida growing: not ornamental, but interesting as a non-European form. In one place, where a small streamlet dribbled down the tufa cliff, Adiantum capillus Veneris in abundance on the dripping rocks. At Praya Formosa, on the West side, fine section of numerous beds of volcanic cinders and tufa sloping at a pretty high angle to the sea, *i. e.* to about S. W. At a distance the beds appear very regular: but on a near view, they are seen to be much waved, and even contorted on a small scale, and very irregular in thickness, thinning out in some places and thickening abruptly in others. At the western point of the little bay, these beds are overlaid unconformably by the ordinary basalt. The abrupt islet rock which lies opposite to this point is likewise of basalt, distinctly prismatic.

Mr. Stoddart, the consul, tells me that the cultivation of the sweet potato in this island has much increased since the vine began to fail. It is a plant of very easy culture, and very productive. He also told me on the authority of Mr. Murray, consul in Teneriffe, that the export of *cochineal*, the produce of Teneriffe, in the first nine months of this (?) year, amounted to the value of £160,000. The rearing of cochineal in Madeira is an experiment newly commenced.

December 28th. 1853.

We rode out along the coast road, to the eastward, as far as the Church of S. Gonzalo, returning
by the Palheiro road. The road as usual, a
succession of steep ascents and descents, crossing
ravine after ravine : fine views of the rugged and
precipitous coast to the eastward, as far as Cape
Garajao, or the Brazen Head : the sea bursting
and foaming gloriously against the black rocks.
The ravines with their mixture of wild rocks and
luxuriant culture, their bold escarpments of black
basalt here and there mantled with fern and
brambles : their terraced plots of soil, rich with
sugar-cane, and orange and fig and peach trees, and
the strange uncouth contorted stems of the prickly
pear, everywhere clambering among the rocks and
walls,—always excite in me fresh admiration. We
saw beautiful masses of china roses, scarlet
pelargoniums and heliotropes, growing half wild
among the rocks. Adjoining the Church of S.
Gonzalo, two fine date-palms form very pleasing
objects in the landscape. A pretty shrubby cassia
in full flower, abundant among the prickly pears
in several places along the coast road, but I
suspect not indigenous. The stone walls in some
places beautifully adorned with Davallia Canariensis
and Polypodium vulgare, both which I saw growing
also, but less luxuriantly in the crevices of basaltic
rocks. A remarkable looking plant, probably Cras-
sulaceous (a Sempervivum) (?) with almost shrubby
stems, and large broad succulent leaves spreading in
rosettes,—reminding me very much of some of the

1853. Cape Crassulas,—frequent on the basaltic rocks, as I have seen it in several other places near Funchal, but without flower. A tallish tree of the Brazilian Araucaria, of its characteristic broad headed form, growing in a garden in the outskirts of the city, as one goes out to the eastward by the coast road.

Fine sections, in various places along this road of strata of tufa and volcanic cinders, covering and covered by thick beds of solid, hard, heavy, dark-grey basalt, in part very compact, in part more or less porous and vesicular. In general, the upper part of each bed of basalt is vesicular, the middle and lower parts very compact. In other respects the character of the basaltic lava appears to be very constant, while those of the different layers of cinders and scoriæ and tufa are extremely various. In one place I observed numerous angular pieces and blocks of basalt imbedded in the tufa.

———

December 29th.

Torrents of rain all the morning with a boisterous S. wind, the rain falling in deluges like those of the tropics. The torrents of Santa Luzia and S Paulo, which pass through the city, were very much swollen, filling the whole breadth of their channel, and came rushing and roaring down with a mighty noise and fury. The sea, however was not nearly so high as on the 27th. Towards 2 p.m. the sky cleared for a while, and the mountains partially emerging from the mist, indistinctly seen in dim

gigantic masses, looked truly grand. A succession 1853.
of beautiful rainbows all the latter part of the afternoon. The evening again very stormy.

FUNCHAL.

The fruits now in season here are bananas, which
are excellent : oranges also very good, especially
the delicious little Tangerines,—apples, guavas,
(Psidium pomiferum) and the anona or custard
apple (Anona squamosa). (?) This last is of an
irregular roundish egg-shape, the outside of a dusky
greenish-brown colour, marked in a somewhat
reticulated pattern with slightly raised lines, which
are the vestiges of the lines of junction of the
different carpels composing the fruit, the interior
(the partitions between the carpels being completely
obliterated), is filled with a delicate white creamy
pulp, of a very agreeable flavour. In this pulp are
imbedded a variable number of large, hard, dark
brown oval seeds, either two-edged or very unequally
three-sided. There are generally some traces of the
central column or receptacle, to which the carpels
were originally attached. The tree which is
frequent in the gardens here, has large, handsome,
smooth, oblong leaves, of a fine slightly glaucous
green, arranged in a very regular two-ranked and
alternate order on the slender branches. Six
different varieties of bananas I am told, are cultivated in this island, of which that called the silver
banana is the finest.

December 31st.

M. Hartung (a German naturalist settled here), tells Lyell that the Papilio (Vanessa) Atalanta found in this island, is a slight variety, differing from the European form by a minute but constant difference in the complicated markings of the under-side of the wings. A similar (but not the same) minute mark distinguishes the North American variety of Atalanta. Mr. Lowe ascertained 60 species of Helix (including Bulimus and Achatina) in Madeira and Porto Santo; Mr. Wollaston has since discovered nearly as many more.

LETTERS.

Funchal.
January 3rd, 1854.

1854.   My Dear Father,

We continue to enjoy Madeira very much, though the weather has not been as favourable as might be; indeed it is considered uncommonly bad weather for this country, very changeable, stormy and at times raw and chilly, with frequent and sometimes very violent showers. Last week indeed for two or three days, it rained in such deluges as to remind me of the tropics, and the little streams that flow in deep channels through the town, were swollen into deep and furious torrents; while the wind blew with such violence from the south, that every vessel was obliged to quit the bay, for the anchorage is so insecure and the bay so much exposed, that all vessels are obliged to stand out to sea 1854. when it begins to blow from the south or south-west, to avoid being driven on shore. Towards the end of this gale arrived the Sierra Leone packet in five days from England. She landed her mails, but as the weather was so rough and threatening, the Captain did not choose either to land his passengers, or to wait to take on board the letters for Teneriffe; but the unfortunate passengers for Madeira were carried on to be landed at the Cape de Verds or at Sierra Leone!

On Saturday, the St. Jean d'Arc, of 101 guns. Captain Keppel* came in; I happened to be on the beach when she came to anchor, and saw her salute the forts, which was a pretty sight. She went to sea again on Monday, Captain Keppel's orders being to cruise in these latitudes. He offered Charles Lyell passage to Porto Santo, which Lyell gladly accepted, as he wanted very much to examine that island, and it is seldom that there is any opportunity of going thither, except in the wretched country boats: so he is gone off on an expedition of some days: Mary remaining here with us.

The weather is still very unsettled; the morning often beautiful, but the clouds generally gather about the mountains early in the day, and gradually descend, so that however fine it may be when we set out on an excursion, we have always a good chance of a wetting before the end of it. The changes of temperature too are frequent and rapid,

* Afterwards Admiral Sir Henry Keppel.

1854. so that it must be trying weather to delicate invalids. But still it is very different indeed from an English winter; on the worst days, there are beautiful gleams, and when the sun does shine in this country, he is sure to make himself really felt; unlike the cold sickly ineffectual mock sunshine of an English winter's day. And then the conflict of rain and sunshine produces more frequent and more splendid rainbows than I ever saw in any other country. I continue to be as much charmed as ever with the scenery and the gardens of Funchal.

It is a constant delight to me to see the beautiful Bignonias, Daturas, Passion-flowers, &c., that hang over the garden walls and cover the trellises and porches. But of native plants there are *very* few in flower at this season, so few as rather to surprise me; not a greater proportion, I think, than at Nice, notwithstanding the considerable difference of latitude. I have as yet collected only fourteen of the forty Ferns which have been discovered in Madeira; the best localities for them are in the interior and higher parts of the Island.

We made a little excursion up the hills yesterday, and had a pleasant botanizing ramble in the chesnut woods, but did not find very much. We returned down the hill, from the Mount Church (Nossa Senhora do Monte) in a very curious conveyance—a basket-sledge,—something like a large basket placed upon skates; we all three sat side by side, and three men impelling and guiding the sledge (one on each side and one behind), we went down the all but perpendicular road at a most astonishing pace. It was an odd sort of locomotion, very like the 1854. accounts of the "Montagnes Russes."

(January 4th). This has been a most beautiful day, quite a summer day, and we made a delightful expedition to what is called the Little Curral, a beautiful wild glen of great depth among the mountains to the north-east of the town. It abounds with Ferns, and I made some additions to my collection. The wild and beautiful scenery, the interesting botany, the bright sunshine and pleasant temperature of the air, our sociable and well-assorted little party, made it a most enjoyable day. Excessive steepness of the mountains, great boldness and variety in their forms, and singularly deep and abrupt ravines, are characteristics of Madeira scenery. But much the boldest and wildest scenery, we are told, is on the north side of the island. On *this* side, above the cultivated region there is perhaps rather a deficiency of wood; no *native* wood at all; some Chesnut woods here and there, and extensive plantations of Pinasters; but on the whole the upper parts of the mountains appear rather bare, though much less so than the Apennines.

The peasantry here are wretchedly poor, but there are not, as is imagined in England, any visible indications of a state of actual famine, nor do any positively shocking appearances of distress meet the eye. The children and old people are generally beggars, but it is difficult to know how far this is the effect of actual want, and how far it is habitual. The people seem very industrious; on the steepest

1854. mountain sides, where they are all but precipitous, one sees little patches of cultivated ground, laboriously terraced and supported by stone walls, in places where one wonders how anybody can get to them to work. The failure of the vines was the heavier blow, because the vintage had become the whole reliance of the people of Madeira, they thought of nothing but the culture of the vine.

Since the grape disease, I understand, they have begun to cultivate a greater variety of things, and to try other kinds of produce, but they are much in want of assistance and encouragement from their Government. It is supposed that sugar and coffee might be successfully grown on a large scale in this island. Certainly both the plants appear to thrive exceedingly here, and many individuals use Coffee grown in their own gardens. If the climate be really hot enough to bring these products to full perfection (which perhaps remains to be seen) the great cheapness of labour here would be an important advantage. Trials are now making with respect to the rearing of cochineal on the cactus; this is becoming a very important article of export from the Canaries, but I am afraid it does not follow that it would succeed equally in Madeira, the climate of which is so much more damp than either of the Canaries or of Mexico.

The last mail from England brought us the important news of Lord Palmerston's resignation, and of the disasters suffered by the Turks in the Black Sea. I am very sorry for them, but the tone of the last articles in *The Times* makes me hope

that a strong public feeling is at last roused in 1854.
England on the subject, and that in spite of Lord
Aberdeen's cowardice or Russian leanings, our
fleet and the French will not be mere spectators
of the struggle between might and right.

I hope you may be by this time in Wales, but if
you should be at Barton, pray give my kind remembrances to the Arthur Herveys, whenever you see
them, and also to Lady Cullum. Pray give my love
to Emily, with thanks for her letter to Fanny, and
believe me ever

Your very affectionate Son,
C. J. F. BUNBURY.

JOURNAL.

January 4th,

We rode to the Little Curral, as it is called,
about north-east of the city. We ascended to N.
Senhora do Monte, then turned to the east, and
after riding a little way, found ourselves in a scene
of striking beauty. We were on the edge of a very
deep and wild glen, at the bottom of which a
beautiful mountain stream, white with foam, wound
its rapid and impetuous course between overhanging
rocks fringed with Fern and Heath; the mountain
sides descending to it in declivities all but precipitous, and rising again far above where we were:
everywhere excessively steep, and often rising in
abrupt craggy masses and walls of rock. No wood,
except here and there a few pines. Higher up the
glen the mountains rise into still bolder steeps and
peaks, in somewhat of an amphitheatrical form.

1854. This glen is the Little Curral, and the stream
which flows down it is that which under the name of
Ribeira de Joao Gomez, passes by the eastern
extremity of the city of Funchal.

Here the rocks abounded with Ferns, and I met
with a very handsome one which I had never before
seen in a wild state—Pteris arguta ; it was growing
plentifully on precipitous dripping rocks ; Cystop-
teris fragilis and Adiantum Capillus Veneris were
abundant where there was a drip of water down the
rocks ; Davallia canariensis far from uncommon on
the rocks, and I found a very fine specimen of Asplenium anceps, Asplenium adiantum nigrum, which is (as
far as I have yet seen), by far the most common Fern
in Madeira, was exceedingly abundant here.   I noticed also Asplenium Filix femina (down by the water
side in the bottom of the glen).   Aspidium aculeatum and Pteris aquilina, but not Blechnum boreale.
The large Sempervivum, very abundant on the
rocks, but not in flower.   In this glen were some
very large Laurel trees (Laurus nobilis) but not far
from houses, and probably planted.   We crossed
the beautiful clear rapid stream by a bridge, and
ascended the opposite mountain by a winding road,
commanding from time to time glorious views of
Funchal and the sea, the lower slopes of the
mountains, and the grand headland of Cabo Girao,
closing the view to the westward.   We could trace
the beautiful winding of the little river glittering
along its deep ravine, till it emerged into the more
open valley near Funchal, and at length entered
the sea.   Continuing to the east we soon after

crossed another very fine ravine, through which 1854.
another clear mountain stream goes dashing down to
join the former one. The left bank of this second
stream was beautifully hung with Ferns, among
which I recognized again the Pteris arguta. The
rocks and shady banks by the wayside were in
many places covered with Mosses, but mostly
barren, Polytrichum nanum however was abundant
in fruit, and here and there Polytrichum juniperinum
or one like it; I think I recognised Hypnum
Illecebrum.

After crossing the Caminho do Meio, we rode
chiefly through Pinewoods (of Pinasters) sometimes
along the open dry mountain sides, till we reached
Palheiro, and returned home by the same road
as on December 21st.

The day was beautiful. The rocks in the
Little Curral swarmed with pretty little lizards;
but I saw no butterflies. As we crossed the
second ravine, a kestrel came gliding by us, very
near.

January 5th.

Pontinha. The bed of black lava on the shore
underlying the yellow tufas, is exceedingly slaggy
and vesicular, with that peculiar harshness to the
touch observable in many of the Etnean lavas.
It is in irregular layers, just such as might be
formed by a succession of flows of melted matter,
one pouring over another in the same general
direction. Some of the layers are more compact
than others. It assumes the various strange and
irregular forms that one sees in the most modern

1854. lavas of Etna and Vesuvius, and close to the
Pontinha it arches over in a remarkable manner,
forming a little cave. This, Lyell thinks, may very
probably be contemporaneous with the lava current
itself, and not hollowed out subsequently by the
waves. A few paces further on (westward) the lava
bed is again somewhat arched, and receding a
little, exposes a bed of tufa *underlying it.*

January 6th.

The beautiful cherry-coloured Oxalis which I
first observed at Palheiro, grows plentifully about
the borders of cultivated fields and edges of roads
in many places near Funchal, as between S. Roque
and the Alegria, near S. Antonio, and between this
latter place and the city. If, as I believe, a Cape
plant, and not originally a native here, it seems at
any rate to be thoroughly established.

January 7th.

Examined the Pico de S. Joao, and the Pico do
Cardo—two of those insulated volcanic cones—cones
of eruption, in Lyell's opinion, which are such
remarkable objects in the neighbourhood of
Funchal. They are rounded cones, somewhat
dome-like; the Pico de S. João, which is the nearest
to the city, is mostly covered either with herbage or
with cultivation, but where its structure appears,
it is composed of very well characterized volcanic
cinders, partly loose, partly aggregated and
cohering more or less firmly. Lyell has observed a
basaltic dyke which escaped me. The Pico do

Cardo, which is the highest, is partly clothed with 1854.
plantations of Pinasters, but great portiorts of its
surface are quite bare, and show a deep red clayey
soil (produced, I suppose, by the decomposition of
the scoriæ and other volcanic ejections) cut by the
action of water into strange gullies. Lower down,
where the side of the hill is cut into an abrupt bank
by the road, we see volcanic cinders and tufa,
containing occasional angular blocks of basalt
interspersed.

Gomphocarpus fruticosus abundant on the hill of
S. João; mostly with only withered remains of
follicles, but I found one plant bearing plenty of
fruit, and a few flowers. I have seen it in some
other places near Funchal, but not in good
condition. It is a Cape plant, but seems to be
thoroughly established here.

———

January 8th.

Captain Grey showed me the Cheilanthus Ma-
derensis (Lowe), growing on a wall in the outskirts
of the city, a little way above the Pico Fort.
Captain Grey says that the Gomphocarpus fruticosus grows apparently wild in Sicily; and that
Oxalis cernua has within the last few years completely established itself in several places in that
island, so as to look like a native.

Mr. Veitch tells me that at his villa in the
mountains the Jardin da Serra, which he says is fully
2,700 feet above the sea, he cultivates Fuchsias for
food for his cattle, which are extremely fond of the

1854. leaves and young branches.   There is so little grass, that the cattle are generally fed in winter on the leaves and twigs of trees or shrubs, and he finds that there is nothing they eat with so much eagerness as the Fuchsia.

January 11th.

From Funchal to Jardin da Serra, Mr. Veitch's house in the mountains; about three hours.

As far as the Socorridos valley, the road the same as that we travelled to Camara de Lobos, from thence we ascend in a N.W. direction, passing the Church of Estreito, of which the elevation is about 1390 feet, above this the vineyards become less predominant and continuous, and some way further we enter thin woods of Chesnut trees, which continue to the Jardin.   Elevation of the Jardin, 2525 feet, according to Major Azevedo's measurement; 2490 according to Fanny's.

A profusion of sweet Violets on the banks and about the roots of the Chesnut trees.

January 12th.

Excursion from Jardin da Serra to the Great Curral and Pico Grande.   A very steep ascent from the house to the edge of the Curral, at first, Chesnut woods, with partial cultivation, then turf with scattered Chesnut trees, and presently after, short dry turf with scattered bushes of stunted Broom. When we reached the edge of the Curral, it was filled with so thick a mist that we could only catch occasional glimpses of an awful depth, of the river 1854. winding at the bottom, of precipices shaggy with evergreens, and of towering cliffs opposite to us.

From hence the road (formidably narrow and rugged), runs northward along a very narrow dividing ridge between the two great ravines of the Curral and Serra d'Agoa ; on either hand a precipitous descent of vast depth with jutting crags, and trees seeming to hang to the sides of the precipices.

The ridge and adjacent mountain sides, clothed chiefly with Erica arborea and scoparia, both in a bushy form and without flower, the former not seen before. Evergreen trees (Laurus) on the cliffs of the Serra d'Agoa. The shrubs and trees loaded with long white hanging tufts of Usnea barbata and and other forms of that genus, intermixed here and there with a large orange-coloured filamentous lichen (an Alectoria canariensis, Ach.), new to me. A large handsome Stereocaulon in great plenty on the rocks. Polytrichum juniperinum and aloides, very common all the way. The moist rocks beautifully carpeted with Lycopodium denticulatum.

On a moist precipice which the road skirted, on the west face of Pico Grande, were many striking plants, but none in flower.

Fine picturesque trees of three species of Laurus, shaggy with Ferns, Mosses and Lichens.

———

January 13th.

From Jardin da Serra to S. Vicente ; a very toil-

1854. some journey, owing to the difficulties of the road and the bad weather. As far as Pico Grande, our way the same as yesterday. We had a fine view of the Curral, which was for a time free from mist ; it is a grand ravine of immense depth and bounded by extremely bold craggy mountains, but bare and certainly inferior in beauty to what we saw afterwards. After passing Pico Grande, the road goes winding for a long way round the head of the great valley of Serra d'Agoa, across a succession of magnificent wooded ravines, and skirting immense precipices, which are superbly clothed with most beautiful and picturesque evergreen trees.

The trees are principally Laurus foetens (*Til.*), Laurus Indica, *(Vinhatico)*, and Laurus Canariensis, The Til is a noble, picturesque tree, wild and free in its growth, with rugged bark and glossy deep green leaves. The Vinhatico very handsome, with larger, longer, and brighter green leaves.

A fine undergrowth of Erica arborea, which here often becomes really a tree, and a very picturesque one, with a trunk of considerable thickness ; the only one I had the opportunity of measuring was 3ft. round, but this was comparatively a small one. Vaccinium padifolium (or Maderense) also formed much of the undergrowth in the latter part of our route—a beautiful shrub, with something of the look of a Myrtle, with glossy bright green leaves, which turn to a fine red in fading.

The old trees most beautifully draped with Ferns, Mosses, and large Lichens. Davallia canariensis particularly luxuriant and beautiful. Polypodium vulgare also in very great abundance on the trees, 1854. and remarkably large.

Pteris arguta in great profusion and beauty and of very large size in some of the ravines. Asplenium anceps and Asplenium adiantum nigrum of very fine growth; and Fanny found Asplenium palmatum, which was new to us. The most common Fern was Aspidium aculeatum. We crossed the main central ridge here called the Encumiada, and descended to S. Vincente in a dreadful rain, which prevented any observations, botanical or otherwise.

January 15th.

S. Vicente to Ponta Delgada.

The village of Vicente, situated in a widish valley between extremely steep mountain ranges, which lower down approach each other so as nearly to close the mouth of the valley. The inn is a mile or more from the sea, and there is a considerable descent from it to the Church. There are Bananas even in this comparatively cold valley, but few and shabby; the orange trees fine and loaded with ripe fruit. Vines climbing high up the trees, of course leafless at this season. Ferns in great profusion and beauty here, indicating the moisture of the climate. Asplenium palmatum in abundance on moist rocks and in crevices of stone walls, even in the village of S. Vicente, close to the inn, and thence all the way down to the mouth of the valley. Adiantum reniforme first seen under an overhanging rock, between the inn and the Church of S. Vicente.

1854.   At the mouth of the S. Vicente river, Davallia Canariensis and Polypodium vulgare were growing actually among the loose rolled stones of the beach, within the very spray of the sea. The coast from the mouth of this river to near Ponta Delgada, strikingly beautiful and bold : grand precipitous cliffs clothed with verdure; beautiful waterfalls formed by the numerous torrents rushing and leaping down the precipices to the sea. One water-fall resembles the Staubach,—the water dissolving as it falls into a waving cloud of fine spray. Another, broken into a succession of leaps in various directions reminds one of the Giessbach.

January 16th.

Ponta Delgada.   Made an excursion to the valley of Bonaventura, but little way in actual distance from Ponta Delgada, but the difficulty of climbing the lofty abrupt headland immediately to the east of Ponta Delgada by a road formidably steep and rugged makes the time required considerable. Myrica Faya plentiful on this headland : a tall handsome evergreen shrub, with fine glossy dark green leaves, so like those of an Arbutus that I took it for one till I found it in flower.   It has no aromatic scent that I can discover.

January 17th.

From Ponta Delgada to Sta. Anna ; a journey of about six hours, taking it leisurely.   We cross first the great headland (already mentioned) between

Ponta Delgada and Bonaventura ; then successively 1854.
two enormous ridges, which branch down from the
high mountains of the interior, inclosing between
them the comparatively low and level plain of the
Arco de S. Jorge, and terminating towards the sea
in stupendous precipices. The road over them
really tremendous. The first or westernmost of
these two last mentioned ridges is peculiarly grand,
wildly serrated with huge crags like gigantic battle-
ments.

After ascending the last of these ridges to the east
of the Arco, we came out on a sort of open plateau
or table-land, comparatively level, and mostly
pasture land, with few trees ; the soil a deep red
clay, much like the red sand-stone districts of
Britain, in parts quite vermillion-coloured. This
table-land extends a good way eastward to within a
short distance of the village of S. Jorge ; there
is then a steepish descent, but still through the
same sort of open clay country to the Church of S.
Jorge, which stands on a smaller plateau, and from
this again a very long and steep descent into the
immensely deep ravine of S. Jorge, which is one of
the finest we have seen. A beautiful waterfall
down the side of the ravine. The river of S. Jorge
very strong and impetuous, rather difficult to cross.

From this ravine we ascended again to the fertile
plateau on which is situated the comfortable inn of
Sta. Anna, 900 or 1,100 ft. above the sea level
(according to the different measurements of the
U.S. expedition and of Captain Vidal). A curious
little green flowered Orchid (according to Mr.

1854. Hartung it is Peristylus cordatus ; Lindley, Haben-
ariæ cordata, Robert Brown) first seen on the rocks
of the headland to west of Arco de S. Jorge, among
Sempervivums, afterwards more plentifully among
Moss and Ferns, on a low stone wall in the plain
of the Arco.

Two large and conspicuous species of Sempervivum exceedingly plentiful on the bare precipitous
rocks of all the ravines and headlands along this
coast ; the one (S. tabulæforme ?) forming with its
broad leaves remarkably flat and compact rosettes ;
the other (S. glutinosum ?) more bushy and straggling, with a stout, woody, branched stem. Several
smaller species of Sempervivum and Sedum in the
same situations, but none in flower.

January 18th.

At Sta. Anna ; a lovely day. Had a very agreeable
ramble through pleasant rural lanes, very like
Devonshire lanes, with steep banks of red earth
overhung with Fern and Brambles, and abounding
with Violets ; and across pretty little mossy and
ferny dells that remind one of Wales and the
north of Devonshire. This plateau of Sta. Anna is
well cultivated, and apparently rich and fertile,
but of a very different character from the neighbourhood of Funchal, nothing tropical in its
aspect. Vines trained up the tall Chesnut trees,
and cultivation under them as in Tuscany.

A noble background of mountains to the south
—the Pico Rinvo, Encumiado Alto, and other high
peaks of the great central ridge ; some snow on

them, but no great masses. The very picturesque 1854.
and remarkable headland of Cape S. Lourenco, the
easternmost point of the island, a fine object, to
the south-east. Porto Santo very distinctly seen in
the distance to north-east.

In this walk, I found for the first time, Woodwardia radicans, growing in plenty on the shady
banks in the lanes and on the margins of streams.
It varies greatly in size. Some specimens here
were magnificent. I gathered one frond that
measured above four feet, independent of its
stalk. In this luxuriant state, gracefully drooping
over the streams, it is a truly beautful Fern.

January 19th.

From Sta. Anna to Funchal; between eight and
nine hours, including an hour's halt at Ribeiro Frio,
and several minor halts.

Our way at first through the same pleasant lanes
as yesterday; then over pleasant, wild, breezy
hills, clothed with Furze in full blossom, Whortleberry (Vaccinium padifolium), and Heath (Erica
scoparia chiefly). The Vaccinium clothed at this
season with beautiful red foliage, giving to the
landscape a strongly marked colouring which is
visible a long way. The Fuchsia coccinea, quite
naturalized among the hills, growing like a native
and forming in one place an extensive thicket,
amidst the Whortleberry and other indigenous
shrubs.

After travelling some way over these hills, we
approach the mountains, of which they are the

1854. underfalls, and cross a succession of nobly wild and deep ravines, which all run down converging to the sea near Fayal.

The Peuha d'Aguia, a most remarkable detached mountain on the coast, singularly abrupt and strongly marked in form,—a very striking object on our left. Astonishing peaks towering above us on the right. Cross the ravine of Metada, one of the grandest of all that we have seen in the island, and after a very long and steep ascent, enter that of Ribeiro Frio.

Then comes an open table-land or mountain plain of some extent, resembling a moor in the North of England, and clothed with nothing but short grass, Moss (chiefly Polytrichum juniperinum) and here and there stunted bushes of the Whortleberry. The surrounding peaks equally bare. I did not see any of the Whortleberry on the actual crest of the pass (the Poizo pass) which is not much less than 5,000 feet above the sea ; it re-appears a few yards below it on the S. side, but extends very little way down.

The Funchal side of the mountains very bare, at the same elevation where the N. side is clothed with luxuriant forests.

LETTERS.

Funchal,
January 22nd, 1854.

My Dear Father,

We are lately returned from a nine days tour in the northern part of the island, in which we

have seen most magnificent scenery, and most 1854.
frightful roads,—I hardly know which deserves the
stronger superlatives.

On the 11th, Mary Lyell, Fanny and I, with the
two maids, set out on horseback, (Charles Lyell
had gone on the day before) and made a three hours
journey to the Jardin da Serra, a house in the
mountains belonging to Mr. Veitch, the former
consul at Madeira, who received us very hospitably.
The Jardin, though still on the southern or Funchal
side of the mountains, is about 2,500 feet above
the sea, and has a climate very different from that
of Funchal. The Vine no longer flourishes at this
elevation, and in the Chesnut woods which surround
the place, the vegetation has quite an European
character. Mr. Veitch cultivates the Tea tree with
great success, and at breakfast he gave us tea
prepared from the shrubs in his own garden, and
very good it was, and of a very delicate flavour:
while at the same time the Coffee came from his
other garden in the town. On the 12th we made a
pretty long excursion into the mountains, but the
mist was so thick that we did not derive much
satisfaction from it. The 13th we sat out from the
Jardin to cross the main dividing range of mountains
to the North side of the island. The party consisted
of Charles and Mary Lyell, Fanny and myself,
Mary's maid and Mrs. Rennie, Major Azevedo, a
Portugese engineer officer, M. Hartung, a German
naturalist,—all on horseback, with a *burriqueiro* or
guide to every horse. Item—two men carrying a
hammock for the ladies to use when fatigued ; item,

1854. two loaded mules and some men carrying instruments and luggage; in all, 22 human beings, eight horses, two mules and a dog! I wish you could have seen our *set out.*

We had a long and fatiguing but very interesting journey across the mountains, through some of the most wildly-beautiful and grand scenery; the road at one time running along a very narrow ridge (a "knife ridge, *cuchilla*," as such places are called in Spanish America), between two enormous ravines with a precipitous descent on either hand; afterwards winding for a long way round the head of a great ravine, along the almost precipitous sides of huge mountains clothed with a primitive forest of Evergreen trees. Nothing could be grander than looking down into the prodigious depth and gloom of these wonderful ravines, amidst the wild, fantastic, contorted trees, that seem to hang to the sides of the precipices. These mountain woods of Madeira are not indeed to be compared to the Brazilian forests for majesty and luxuriance of growth, but they are singularly picturesque and striking. The trees are principally three species of Laurel (Laurus, the same genus with the Sweet Bay), with very handsome, glossy, bright green foliage; and there is a rich undergrowth of Heath (Erica arborea and scoparia) the Madeira Whortleberry (Vaccinium padifolium) and other shrubs, mixed with luxuriant Ferns. The Erica arborea indeed grows really into a tree, with a trunk of very respectable size; the only one I was able to measure was three feet in circumference, and this was by no means one of the

largest. The conclusion of this day was not 1854.
agreeable. It came on to rain violently before we
had crossed the dividing ridge ; we were enveloped
in thick mist. The road was horrible, and for the
last two hours and more we were descending by
these break-neck ways through a tempest of wind
and rain, and latterly in the dark, so that it is really
a mercy that we all got to our night's quarters with
unbroken bones. The inn at St. Vincente was
clean, but otherwise a comfortless place enough,
and there was no possibility of getting any fire, either
to warm ourselves or to dry our wet clothes. It is
wonderful I think, that we all escaped rheumatism.
The next day, the 14th, we remained in our cold
quarters at S. Vincente, and it rained and blew
furiously all day.

Lyell, who is not easily daunted, made his way to
a fossiliferous bed on the mountain above, which he
was anxious to examine. For my part I must
confess that I stayed within doors, the rather as I
had no dry clothing to spare. On the 15th, the
weather and our prospects began to brighten, and
we had two very satisfactory and interesting day's
journeys along the north coast from St. Vincente
eastward to Sta. Anna. The coast scenery is very
grand. The numerous ridges which run down to
the coast from the main central range of mountains
all terminate towards the sea in immense precipices,
shaggy with Evergreens, and furrowed by innumerable streams, while these ridges are separated from
one another by deep and wildly beautiful ravines.
The streams that rush down from the summits of

1854. the cliffs to the sea form a variety of beautiful waterfalls ; we saw one strikingly like the Staubach and I should think fully equal to it in height ; the water dissolving as it fell into a waving cloud of fine spray ; and another reminded us of the Giessbach, the stream being broken into a succession of leaps in various directions. The roads along these precipices, and up and down the sides of the ravines are certainly frightful enough—worse I think than any I saw in Brazil.

At Ponta Delgada, between S. Vincente and Sta. Anna, we spent two nights very comfortably at a house kept by the Vicar of the parish. He does not keep an *inn*, of course not, but he extends hospitality to travellers—for money. The arrangement may not be strictly economical, but it is very convenient to both parties.

The inn at Sta. Anna, where we spent a day is extremely comfortable, with a very pretty garden, and situated in a lonely country. Indeed the country round about Sta. Anna is the most enjoyable I have seen in the island, especially for walking ; one is not hemmed in by those everlasting, hateful stone walls, which are one's plague in the neighbourhood of Funchal ; one rambles through pleasant lanes very like Devonshire lanes, between banks overhung with Fern and Broom and Brambles and abounding with Violets, and through little mossy and Ferny dells that remind one of Wales and the north of Devonshire.

The aspect of the country is quite unlike this side af the island, and has nothing of the tropical

character which is so conspicuous here ; perhaps it 1854.
is on the whole more like the south-west of France
than any other country I have seen, but far more
beautiful. The day we spent there (the 18th) was
one of the most delicious I have felt, neither too
warm, or too cold ; soft, yet fresh and exhilarating,
and we enjoyed it thoroughly.

On the 19th (another lovely day) we turned our
faces southwards, crossed the beautiful wooded
mountains, and after a long but interesting day's
journey, arrived just before nightfall, safe and sound
at our lodgings here. We all came back in high
health ; Fanny seems all the better for travelling,
and it is delightful to me to see how much she has
gained in strength, that she can be on horseback
for six or seven hours, on the roughest of roads,
without being at all the worse for it.

The north side of the island is the country for
Ferns, we were quite charmed with their profusion
and beauty there. I collected nine species that
I had not seen before, and most of them interesting
and strongly marked forms—in particular a very
beautiful and singular Adiantum (reniforme), the
Asplenium palmatum, resembling an Ivy leaf in
shape, and the Woodwardia radicans, which is a
grand Fern, with its large gracefully-drooping
fronds, five or six feet long. Altogether from the
wetter climate and especially from the less extent of
cultivation, the northern side of Madeira is certainly
more favourable to botany than this one ; though
here too, several interesting plants are showing
themselves as the season advances. There are

1854. already decided symptoms of Spring; the Fig trees are bursting into leaf, and flowers and young leaves showing themselves on the Oaks (our common Oak cultivated here), butterflies are becoming more numerous) and—a less agreeable indication of the season—the horse-flies are more than ever troublesome. They call this a bad winter for Madeira, it seems to me that it would make a very tolerable English *summer*.

(January 25*)*. It is hardly possible to believe that we are in the month of January.   If it were not for the leafless state of the Vines, and of the Planes and Chesnut trees, one could never persuade one's self that it was winter.   I continually congratulate myself on our having come to this lovely island.   What a tremendous winter you seem to have in England.

We have formed a very agreeable acquaintance with Captain and Mrs. Grey,* who are spending the winter here : *he* is a brother of the present Lord Grey, and of my college friend John Grey ; and *she* is granddaughter of Lady Dacre.   Both have remarkably pleasant, frank manners, and both are zealous botanists, which draws us together very much.   It is a great pleasure to find persons of pursuits so congenial to one's own.—I have no time to tell you of Captain Keppel and the gaieties of the S. Jean d'Arc.   We hear now that Henry Codrington is likely to touch here with his great ship the Royal George.

Pray give my love to Emily and Cecilia and Henry.

Believe me ever your affectionate Son,

C. J. F. BUNBURY.

* Admiral the Hon. Sir Frederick and Lady Grey.

Funchal, Madeira.
January 27th, 1854.

My Dear Katharine,

I rather think I owe you a letter, at any
rate it is a long time since I have written to you,
and I am indebted to you for some interesting
notices of the botany of Simla, especially of the
Ferns. I am disposed to send you some account of
the botany of this most beautiful and charming
island, where we have been spending six delightful
weeks, happily free from the gloom and frost and
snow of the winter in England. I know that Fanny
has written to you since we have been here, and has
therefore given you a general notion of what is
remarkable about Funchal, and our mode of life
here; and by this mail she is sending to your
Husband a capital account of our interesting tour in
the north of the island; you are not therefore to
expect from me in this letter much besides botany,
and I trust to your love of the science for not
thinking me tiresome. The first thing that strikes
a botanist on landing at Funchal is the thoroughly
tropical character of the cultivated vegetation. The
beautiful large, silky green leaves of the Banana
waving in every garden, the Coffee shrub with its
bright glossy foliage, the Custard Apple, the Guava,
and the Sugar-cane; the long white bells of the
Datura arborea hanging over the garden walls;
the most beautiful festoons of Bignonia venusta;
of sky-blue Ipomœas, and of Passion-flowers mant-
ling the walls and trellises and porches of the
houses; such are the objects which meet the eye

every minute as one passes through the streets of Funchal and along those paved roads which extend so far on every side of it. The aspect of the gardens of the city, as one looks over them from any of the higher points of it, is exquisitely rich and beautiful,—quite Eden-like. The climate is so fine that almost all the productions of the Tropics may be raised with ease and will flourish in the open air. Several of the gardens, in which some pains are taken to collect exotic plants, are exceedingly rich, and with a well-directed expenditure in procuring plants, a wonderful botanic garden might be formed. Not only in the gardens, but in the plots of cultivated ground among the rocks, one sees everywhere the Banana, the Sugar-cane, and the Cactus, and frequently the Coffee-bush intermixed with the Fig, the Almond, and the Peach tree. Bananas are a standing dish at dessert, and of very good quality; and frequently we have the Brazilian Anona or Custard Apple, a delicious fruit. Very good Coffee is made from the berry grown here, and though the culture has not yet been attempted on any large scale, it is not at all unlikely that it might succeed, though perhaps only in a limited district. I have omitted to mention the Orange tree, which is universal in the gardens here, and produces very fine fruit. Amidst all these tropical forms, the tall reed, Arundo Donax, which is largely cultivated, reminds one of Italy. Date palms are not as numerous as I should have expected; there are a few very fine ones, but they are not generally conspicuous in the landscape, nor, I believe do they ripen their fruit thoroughly. There are a few Dragon trees (Dracena 1854. Draco) in the neighbourhood of Funchal, but small ones, I fancy, in comparison with what we shall see in Teneriffe: strange uncouth trees they are, with their gouty-looking stems, their stiff, bare branches that thicken upwards, and their scanty narrow foliage, growing in tufts at the very tips. It is a tree that looks as if it might have been contemporary with the Iguanodon and the Pelorosaurus, and quite in keeping with them. So much for the cultivated vegetation of this part of the island, and most beautiful it is. With a view to the native botany, Funchal is rather a disadvantageous station, though not as bad as I at first thought.

Cultivation extends so far, and has taken such complete possession of the soil, that one must go a good way before one meets with any of the wild natural growth of the island, and indeed it is long before one can get quit of the wearisome paved roads, and still more wearisome stone walls. The greater number of the plants that one meets with in the immediate neighbourhood of Funchal are either such as come under the denomination of *weeds*,—plants that grow at the edges of roads and in the borders of cultivated fields,—or else naturalized exotics. Indeed, foreign plants establish themselves so readily here, that it is difficult and becoming more and more difficult to discriminate between the native and the introduced vegetation. For example, a beautiful little cherry-coloured Oxalis (either speciosa or humilis), a native of the Cape, which was first introduced into the Island by

1854. a lady now living here, is become quite wild and very abundant in many places, and has quite the appearance of a native. So also some of the Cape Pelargoniums may often be seen growing apparently wild.

The Fuchsia coccinea and Datura arborea are rapidly establishing themselves; on the northern side of the island, where the climate is moister, they form large thickets in some of the ravines, mixing with the natural growth, and in twenty or thirty years more, I dare say they will be well-established wild plants of Maderia. The common Broom— Spartium scoparium,—which is one of the most common shrubs on uncultivated ground, and covers large spaces even on some of the high mountains, is said to have been purposely introduced. I have little doubt that many of the common Madeira plants, whose introduction is less recent and not recorded, are in fact immigrants through the unintentional agency of man.

Still there are some places in the neighbourhood of Funchal, where the truly indigenous vegetation may be studied. One of the best is a deep glen commonly called the Little Curral, about an hour's ride from hence: it is a very picturesque spot, little intruded upon by cultivation, the rocky banks of the stream abound with Ferns, and in particular two beautiful and interesting species of that family are to be found there,—Adiantum reniforme and Woodwardia radicans. It so happened, however, very oddly, that I did not find these two Ferns in the Little Curral, till after I had seen them plentifully

on the northern side of the island, where they are
much more abundant than on this side.   In the
same glen, too, there still remain some scattered
trees of the beautiful evergreen *Vinhatico*, Laurus
Indica, the relics of the forest which once clothed
these mountains.

On the sea cliffs, too, to the eastward of the city,
some interesting plants are to be found, though not
many are in flower at this season.   A magnificent
shrubby Echium is just come into blossom, and the
fine warm weather that we have had lately calls out
new flowers almost every day.

The Vine cultivation does not extend higher in
general than about 2,000 feet ; above this there are
some woods of Chesnuts and Pinasters (planted),
but in general the mountains are very bare on this
the southern side.

When we penetrate into the interior and cross the
watershed of the mountains towards the north
coast, the scene is very different : the steep sides of
the mountains and the deep and wild ravines which
intersect them, are everywhere clothed with a rich
forest of beautiful and picturesque evergreen trees,
principally Laurels of four species : the Laurus
foetens (or Til); Laurus Indica (Vinhatica); Laurus
Canariensis, and Laurus Barbasano.   The undergrowth consists of two Heaths,—Erica arborea and
scoparia, growing to a great size,—of the Madeira
Whortleberry (Vaccinium padifolium of Smith, a
beautiful shrub), and a few others.   The Erica
arborea grows really to a tree and a very picturesque
one, with a gnarled and twisted trunk of consider-

1854. able bulk. All the old trees in these forests are most beautifully draped with Ferns (especially Davallia Canariensis and Polypodium vulgare), Mosses and Lichens. I collected a good many interesting things belonging to these two latter families, but owing to the length of the days' journeys we were a good deal hurried both times that we crossed the mountains, and my attention was so much engrossed by the beautiful Ferns, that I could not spare much time for the more minute cryptogams.

The north side of the island, owing to its wetter climate, and milder condition, is the true country for Ferns. There profusion and beauty are quite enchanting. In all the deep and wet ravines which intersect that coast, we see them clothing the rocks and shady banks, the roots and trunks of trees, the margins of the innumerable rushing streams, and the moist mossy walls: even in the villages they abound. The Asplenium palmatum, Asplenium Canariensis, Asplenium monanthemum, and Adiantum reniforme, elsewhere rare, are abundant along that coast, and Woodwardia radicans, and Pteris arguta grow to magnificent size and beauty. The beautiful Hare's foot Fern, Davallia Canariensis, and the Polypodium vulgare, are the two most generally common Ferns throughout the island; they are frequent even in the immediate neighbourhood of Funchal, and on the northern side there is hardly a moist wall or an old mossy tree on which they do not flourish; I found them growing even on the actual beach.—I have during the time we

have been here collected 25 out of the 40 Ferns 1854.
which are known as natives of Madeira, and I
hope to get several more before we leave it. I will
keep specimens for you of as many as I can : as I
think it may interest you. I enclose a list of all
the Ferns and flowering plants I have hitherto
collected in Madeira, from the 18th of December to
the present time. The number of flowering plants
may appear small, if you do not consider the
season : but in spite of the warmth of the Madeira
winter, a very small proportion of the native plants
flower at this time of year.

This is certainly a most lovely island ; the two
sides of it are very different, but both extremely
beautiful. We had a delightful tour in spite of
frightful roads, and of a day-and-a-half of very bad
weather. We all came back in high health and
spirits, Mrs. Rennie and all. Fanny is delightfully
well and strong, full of spirit, " up to anything," is
on horseback for several hours a day, and quite a
different creature from what she was a year ago.
She has made a great number of very characteristic
and accurate sketches, both here and in our tour,
which give an excellent idea of the forms and
grouping of the mountains, and of the outlines of
the scenery ; and she is indefatigable in making and
calculating barometrical observations for Charles
Lyell. The fine weather, and being much in the
air, agrees admirably with me, and I thoroughly
enjoy Madeira.

Charles Lyell is activity personified, and has
made great discoveries in the geology of the island.

1854. I have not strength or activity enough to keep up with him, so Mary and Fanny and I generally keep together in our excursions.

You are no doubt now thinking much of Leonora's approaching marriage; it offers, I trust, every prospect of happiness; her situation at Berlin will be a very agreeable and satisfactory one for a person of such literary tastes and acquirements, and all we hear of Dr. Pertz is so agreeable that I am anxious to be acquainted with him.

I wish we may see you in England this year, but I can quite conceive what difficulties and uncertainties you must have in deciding what is best to do. Pray give my love to your Husband, and believe me,

Your very affectionate Brother,
C. J. F. Bunbury.

Madeira,
February 3rd, 1854,

My Dear Leonora,

This is a most charming island, more beautiful in point of scenery than anything I ever saw before, except the neighbourhood of Rio de Janeiro, and delightful, though a little variable and uncertain in point of climate. Its botany too is rich and interesting.

Of the plants of cultivated ground (agriculturally speaking, *weeds*) a great many are now flowering; yesterday we found 14 or 15 that we had not seen before, and many of them very beautiful; but all these are plants of the south of Europe, whereas of the peculiar Madeira plants very few flower at this season. These latter (I mean plants which are either absolutely peculiar to Madeira, or common to it, with the Canaries, but not found on the continents of Europe or Africa) are mostly plants of the sea cliffs or of the woods and mountains of the interior. One of them which is now in blossom, a magnificent shrubby Echium with a huge thick spike of light-blue flowers, is popularly called the Pride of Madeira. We went to see a curious rock, yesterday. It is a vertical chasm communicating with the sea, in a headland a few miles west of this. One descends into a hollow, much like a small crater, enclosed by the wildest rocks of black scoriaceous lava, and in the middle of it, amidst rocks which are tumbled about as if giants had thrown them together in sport, opens a grim chasm, at the bottom of which, at the depth of 47 feet, is the sea, rushing in through a natural tunnel. Lyell measured the depth by means of a measuring tape with a stone fastened to the end of it. The grim blackness and extraordinary shapes of the lava rocks around, are the most striking features of the scene.

Believe me ever,

Your very affectionate Brother,

C. J. F. BUNBURY.

JOURNAL.

Madeira,
February 9th.

Excursion to the Curral das Freiras, or Great

1854. Curral.   Started from Funchal about half-past
eight ; rode at first in a N. W. direction, past S.
Amaro and the Pico de S. Antonio and Pico do
Cardo.

After passing some pine woods, we came out on
the side of a bare mountain ; in front of us a
mountain rising into a very bold, nearly conical
peak, crowned with a very remarkable tower-like
mass of rocks of distinctly columnar basalt. The
path goes winding round the head of a wild little
valley tributary to the Socorridos, along the grassy
sides of bare mountains, not very abrupt, the whole
resembling a mountain scene in Wales or Cum-
berland. After rounding the shoulder of the mountain with the crest of basaltic crags, the scene
changed at once ; we were on the edge of a great
precipice, looking down into the fearfully deep and
gloomy ravine, or rather chasm, through which the
Socorridos flows from the Curral ; opposite to us the
mountains from which we had looked down on the
same gorge on the 12th of last month ; and further
to the north, full in view, the gigantic cliffs and
craggy summit of Pico Grande. Very near to
the grassy brow from which we looked into the
Curral (Monte Grande I understand it is called),
to the right of the road, one of the most striking
basaltic dykes I have ever seen rises abruptly
out of the mountain slope. It is a perfect wall
of rock, not more than eight or ten feet thick,
quite vertical on both sides, rising to the height
of twenty or thirty feet on one side, and of forty
or fifty on the other (for the level of the surface

from which it rises is much lower on the one 1854.
side), and running straight down the steep slope of
the mountain for a considerable distance.   I saw
no appearance of the spheroidal basalt near the
Curral.   I saw no plants in flower except the pretty
little Thyme, like Micromeria variæ, and one bush
of a handsome kind of Wall-flower, Cheiranthus
mutabilis

February 11th.

Ribeiro Frio.

In the beautiful evergreen woods of that valley
and the Metade, the most common tree appears to
be the Laurus Canariensis, now coming into
flower, and it grows to a very large size.   Some of
the hollow and shattered trunks appeared to me
to be as large as most English Oaks.   I recognized
also the three other Laurels, — Laurus fœtens,
Indica and Barbusana,—this last well-marked by
its hollow leaves, like an inverted spoon: and the
Clethra arborea, or Folhado, a tree with beautiful
large bright-green leaves.   Along the Levada, or
water-course called Levada do Forado, which winds
round the mountain-side from the Ribeiro Frio to the
Metada ravine; the vegetation is exceeding rich and
beautiful.   Ferns in the greatest profusion, remarkably fine and luxuriant.   Trichomanes speciosum
and Hymenophyllum tunbrigense here and there on
the wettest parts of the rocky bank, in the deep-
est shade, among moss and the creeping Sibthorpia,
often under overhanging stones and roots of trees:
the Trichomanes especially in the deep cuttings,

3 P

1854.   where very little light enters. The Hymenophyllum rather plentiful in some spots, but not in anything like the same profusion as at Killarney, and scarcely as large as *there*. Acrostichum squamosum at the roots of decayed trees along this Levada, in very shady places : very rare.

The view of the Metade valley, and the mountains at its head from the Levada do Forado, is the most beautiful and wonderful I have seen in the island. Looking upwards we have a glorious range of the wildest and most astonishing crags and pinnacles I ever saw, including the highest peaks in the island, on the other hand looking down into the deep valley of the Metade, we follow its windings between the precipitous mountains, shaggy with rich evergreen wood, till it is closed by the bold mass of the Penha d'Aguia, and the sea beyond.

LETTERS.

Funchal,
February 15th, 1854.

My Dear Father,

I must write to you before we leave Madeira.

We mean to go to Teneriffe by the next Brazil packet, which will probably be here the day after to-morrow. I shall leave Madeira with regret, and shall always look back with very great pleasure to our two months stay. I have enjoyed it very much, though I own I should have enjoyed it still more, if we had not lived in such a perpetual hurry and

excitement all the time we have been here, which 1854.
does not exactly suit my temperament. As for
Lyell, I do not think that, except when asleep, he
has been stationary anywhere for two hours, since
we arrived. His activity and spirit of adventure
are the wonder and admiration of everybody. He
has made great discoveries, and will by-and-bye, I
have no doubt, produce a first-rate memoir on
Madeira.

It was a very great pleasure to me to receive
your letter of the 20th and 22nd of January, and I
thank you very much for it. It so happened that
our batch of letters, all very satisfactory, arrived on
my birthday, just as Fanny and I returned from a
very pleasant excursion, and very happily was the
day wound up by reading them. Your apprehensions
that the climate of Madeira would be too relaxing
for Fanny's health have been shown by the event to
be unfounded, for I have never seen her so well
or so strong or in such constant good spirits as she
has been during the whole of our stay. It certainly
does seem odd that so damp a climate should
agree so well with her, for it is so constantly moist,
that the drying plants is a slow and tedious
operation. She is indefatigably active, — observes
the barometer and works out the calculations of
heights for Charles Lyell, helps me in drying plants,
and has made a great number of very accurate
drawings of the scenery, and especially of the forms
of the mountains, many of which are indeed
very extraordinary, and not to be described in
words.

1854. *(February 16th).* The steamer is signalled a day sooner than we expected, so I can only add a very little more. We have made some delightful excursions, in particular one last Saturday, in company with the Greys and Mr. Baring,* to Ribeiro Frio, a ravine on the north side of the mountains, among those beautiful evergreen Laurel woods which I described before. We had a most agreeable day in every respect. The scenery we enjoyed was the finest I have seen in the island, and the vegetation especially the Ferns, remarkably luxuriant and beautiful.

We hear a sad account of the condition and prospects of the poor people of this island. Bread and all kinds of provisions are rising excessively in price, there is no employment for the people, and actual famine is apprehended. When the season comes that the greatest part of the foreign visitors depart, and the resources derived from their expenditure are withdrawn, there seems to be nothing in prospect for the unfortunate people but starvation. The government does nothing : it even lays a tax on emigration, which it ought to encourage, and there seems to be no class in the island sufficiently well-off to assist others ; for the landed proprietors have suffered so heavily from the failure of the Vine, that, as we are told, families previously wealthy are selling their furniture and their silver spoons to procure subsistence. It is Ireland over again, or worse. No country seems to be more in need of an " Encumbered Estates Bill ;" most of

---

* Afterwards Bishop of Durham.

the landed properties, it is said, are strictly entailed, 1854. and heavily burdened with provisions for younger children, and all kinds of encumbrances, and a most complicated system of law. It is sad to think of such misery impending over the inoffensive people of this beautiful island. My impression, from what I have seen of the common people of this country is very favourable. I never met with a more civil, obliging, good-humoured, harmless people. Thefts of course there may be, but I should think one could nowhere feel more secure from violence or outrage of any sort.

We are still in the dark as to whether it is to be peace or war. The latest positive information we have is, that the Russian ambassador had left London, and that Lord Dundonald was to be sent with a fleet to the Baltic,—indeed I do not know whether this last part is certain, or only a report. This looks warlike. Just before, we had seen newspapers with Cobden's and Bright's speechifying at Manchester, and I was apprehensive that, backed by that wretched Manchester party, Lord Aberdeen would contrive after all to temporize and palaver till the Czar had done his worst with Turkey. I do not feel so much curiosity about the Queen's Speech, which I have no doubt will manage to say as little as possible, as I do to hear what our fleet has been doing. By the way the vessel which was mistaken for our steamer this morning, has brought a report, — but it seems to be only a report, — that part of the Russian fleet in the Black Sea has been destroyed by the

1854. English and French fleets. Well, if the war is begun, as I think it ought to be, I hope as you say that the English and French governments will do their work thoroughly, and not dally with the war,—not manage it as England did the last American war.

We have now in our stay of less than two months, collected thirty-four Ferns, out of the forty known in Madeira.—I like Captain and Mrs. Grey more and more, they are capital botanists, full of zeal, and very accurate. I never new anybody with so quick an eye for plants as Mrs. Grey has.

The Aubrey Spring Rices too I like very much. I feel really sorry to leave Madeira, and should be still more sorry to think that I should never see it again. I certainly do not expect ever to see anything more beautiful. I should like of all things and so would Fanny, to spend a month here on our way back from the Canaries, but Leonora's marriage will probably induce us to return to England in April,—which will be returning from summer to winter.

I have no time at present to write any of my thoughts upon politics, but I will write to you from Teneriffe. Pray give my love to Emily, who I trust is well, and also to Cecilia and Henry, and to Sally and her little ones, if this finds you at Barton.

Believe me ever,

Your very affectionate Son,

C. J. F. BUNBURY.

P.S.—Our host (who is secretary to the muni-

cipality of Funchal), and his sisters have been uni- 1854.
formly most obliging and good-natured, and have
made us extremely comfortable during the whole of
our two months stay in their house. Certainly they
are calculated to give us a most favourable idea of
the Madeirans.

Santa Cruz, Teneriffe,
February 20th, 1854.

My Dear Mr. Horner,

  I have long been intending to write to you;
and now, being in quarantine, and therefore having
some leisure, I will really make an attempt.—
  I hardly think I ever in my life passed two months
more pleasantly than the two we have just spent in
that most beautiful and charming Island of Madeira.
  We thoroughly enjoyed our long and interesting
exploring rides, sometimes by ourselves, sometimes
in company with our pleasant friends the Greys.
  What added still more to my enjoyment of
Madeira, was the happiness of seeing my dear
Fanny improve so marvellously in health and
strength. She has been making a great number of very accurate and interesting drawings of
the scenery, but what drawings must fail in giving
any tolerable idea of, are the luxuriant beauty and
almost tropical richness of the cultivated country
about Funchal, glowing with Brazilan and Indian
flowers, and with the splendid foliage and bright
tints of the Banana and Palm and Orange and
Sugar-cane ; the wild grandeur of the primitive
Laurel forests on the central mountain chain.

1854. The absolute number of Ferns in Madeira is pretty nearly the same as in the British Islands, but their number *relatively* to the flowering plants about twice as great in the former as in the latter.

We had a quick and pretty comfortable passage of only twenty-seven hours from Madeira hither, but it was great good luck that we did not lose our luggage and get a thorough ducking into the bargain; in landing here, indeed, if it had not been day-light we might have been in real danger.

I am glad to see in the Queen's speech announcements of several important internal reforms, *not* including Parliamentary Reform, which last I cannot help considering as somewhat of a humbug. If, however, anything of that kind is wanted, a time of so much internal tranquility is no doubt a good one for bringing it forward, as I hold that the best time for proposing any reform is when there is not a clamour for it.

Give my affectionate love to all your party.

Ever your affectionate Son-in-law,

C. J. F. BUNBURY.

———

Santa Cruz, Teneriffe,
March 4th, 1854.

My Dear Father,

The mail has this morning arrived from England, and we have received our letters and have had the comfort of hearing that all is well with those we are most interested in, thank God.

With respect to public affairs, it seems astonishing that we are not yet avowedly at war, but I read with

great interest and satisfaction of the preparations 1854.
making for carrying on the war in good earnest
by land as well as by sea. If the accounts in the
newspapers are to be trusted, there is a noble fleet
preparing for the Baltic, and our contingent of
troops is likely to do us credit. I hope all this is an
indication that the war which seems inevitable, and
which our ministers have certainly done all in
human power to avoid, will be carried on with
vigour, and that the gigantic prestige of Russian
power will be effectually broken, and securities taken
against further encroachments. Our soldiers as
well as sailors seem to be quite ready and full of
spirit, and I conceive that both services are in a
very much better state than at the beginning of the
last war. But after all, however much one may
approve of the cause, a great war such as this will
be is a serious evil ; interrupting the steady course
of internal improvement, interrupting the progress
of science and literature and all peaceful pursuits,
adding largely to the burdens of a country already
so heavily burdened as ours, and operating un-
favourably in so many ways on the condition of the
mass of the people. I have no hesitation in wishing
for a war rather than any other alternative that
seems now probable, or in agreeing with you that
the preservation of peace is even (under the circumstances) undesirable : but I conceive that we must
be prepared for much that will be disagreeable, and
that the popularity of the war will not last very long
after people begin to experience its effects.

My last letter to you was written just before

1854. we left Madeira. We were kept only three days in quarantine, and were much more comfortable than I had expected, for as the lazaretto happened to be full of passengers from Havana, Mr. Murray, the Consul, had got leave for us to remain in a separate house, and we were supplied with furniture and food from the hotel. The weather was cold and our rooms rather too airy, but otherwise we were pretty well off, and we had so much to do, that really I was half sorry our detention was so soon over!

The very morning after our release the Lyells crossed over to Grand Canary, and are still there; but I was inclined to see something more of this neighbourhood, and have found much of interest in it. The character of the scenery is quite different from that of Funchal, and certainly much inferior in beauty; on a general view it appears bare and savage, and even dreary, but some of the little valleys concealed behind the rugged coast mountains are really beautiful when one comes to explore them, with charming little spots of rich cultivation, —tufts of Orange and Peach trees and Bananas and Palms,—scattered amidst the wildest rocks. The native vegetation is extremely peculiar and interesting. One of the most abundant plants is a large, succulent, leafless, angular, prickly Euphorbia, a most thoroughly African type, just like some of those which prevail in the eastern part of the Cape colony, and which you may see in my book. It looked to me quite like an old acquaintance. It grows in vast quantities all over the rocky hills and sea cliffs along this coast in large clumps, dotted

about the hills and rocks, so as to give them a 1854.
singularly spotty appearance, that is conspicuous
even from a distance. Another characteristic plant
is the *Balo* or Plocama pendula, a shrub with long
slender weeping branches and still more slender
leaves, the whole aspect of the plant peculiarly
light and gracefully pendulous, like a miniature
weeping willow ; this is peculiar to the Canaries,
and is as abundant as the Euphorbia. There are
also some other shrubs of a succulent nature and a
peculiar character ; a great variety of Houseleeks ;
two beautiful species of Lavender ; *shrubby* Sow-
thistles ; and a great variety of curious plants
besides. I have had ample occupation in the
botanical way during our stay here, and have no
doubt I should find enough for a long time. But
as we do not wish our knowledge of Teneriffe to be
confined to Santa Cruz, we are going to start the
day after to-morrow for Orotava, on the opposite
side of the island, celebrated for its beauty. We
shall there have at least a good view of the Peak,
though I suspect the ascent of it will be impracticable at this season. It is a very inconspicuous
object in the view from hence.

The Consul and Mrs. Murray are extremely good-
natured and friendly to us, and so are Mr. and
Mrs. Hamilton : and I have got acquainted also
with M. Berthollet, the French Consul, a distinguished botanist, one of the authors of a splendid
work on the natural history of the Canaries ; he
is very civil to me, and seems much pleased to find
a botanist to talk with. But it is seldom I meet

1854. with people whom I take such a fancy to, as to Captain and Mrs. Grey, whom we saw so much of at Madeira.

I am rather anxious to know how the Arboretum has borne the severe winter that you have had, and whether any of the more valuable plants have been killed or seriously hurt; some of them I should think have hardly been exposed to such a trial since their first introduction into England. I hope sufficient precautions were taken in time. I hope you and Emily are now both well. I am very glad to hear that your book, which I have read with such delight, is to be made more generally known.

> Believe me ever,
> Your very affectionate Son,
> C. J. F. BUNBURY.

————

> Puerto de Orotava,
> Teneriffe,
> March 14th, 1854.

My Dear Father,

My last letter to you was written from Santa Cruz, on the 4th. We set off on the 6th and had a successful journey hither, without any accident, though we were rather fatigued when we arrived. The distance is called seven leagues, which ought to be only twenty-one miles, but the reckoning of distances in this country is very uncertain, and this is reckoned by the English residents at twenty-four or twenty-five miles, and is probably not less. For the first two leagues from Santa Cruz there is a

tolerably good carriage road, passing through the 1854.
city of Laguna, the old capital of the island, a
decayed and gloomy town, not very interesting. At
the end of the carriageable road we were met by a
hammock which a gentleman of this place had
kindly sent for Fanny's accommodation, and she
travelled part of the way in this, and part on a
donkey. I rode a pony which the Consul had lent
me. The Lyells, I should have mentioned, were not
yet returned from Canary. We had a most beautiful,
bright, warm day for our journey, and enjoyed a
splendid view of the Peak, which looked all the
grander for being covered very far down with snow.
From about the middle of the way the view of it was
especially glorious, the lower ridges half lost in
the hot bright haze, and a girdle of light clouds
floating about its middle, while the cone with its
dazzling snow seemed to soar to an immeasurable
height into the sky. It is indeed a noble mountain.
We have spent a week very agreeably in this
beautiful place, enjoying delicious weather—weather
indeed "*fait à souhait*" for it is neither too hot
nor too cold.

The valley of Oratava, which has an immense
reputation, and is called the Paradise of the
Canary Islands, is indeed very lovely. I do not
however think it *more* beautiful than Madeira, as
the people here boast, indeed both are so charming
that it would be invidious and useless to exalt the
one above the other. This valley however is said to
have lost some of its beauty of late years, through
the destruction of much wood, and the substitution

1854. to a great extent of the culture of the Cactus for that
of the Vine.

The rearing of Cochineal is now the most
profitable branch of industry in these islands,
and certainly it does not contribute to the em-
bellishment of the country, for the Prickly Pear is
as little beautiful as any plant can well be.

We are tolerably well lodged here, and from the
flat roof of our inn we have a noble view of the
valley, the mountains, the sea, and at times, of the
Peak. This last is very commonly hidden by clouds
during great part of the day, but in the morning,
and still more in the clear, moonlight nights, it is a
glorious object. By moonlight indeed it is indescribably beautiful, when its regular, snowy pyramid,
so brilliantly white, is seen relieved against the
clear sky, and the long wall of lower mountains
over which it towers is in deep shadow. Nothing
can be more lovely than the nights in this climate.

We have a very agreeable acquaintance in a
gentleman of the name of Smith, who has lived here
many years, and is most intimately acquainted
with this country and everything belonging to it.
He took high honours at Cambridge, and is a
man of great scientific attainments, and of highly
cultivated mind in many ways, and remarkably
gentleman-like and agreeable, really a wonderful
man to meet with in this remote corner of the
world. I cannot say enough of his kindness and
attention to us. From the British Vice-Consul,
Mr. Goodall, also, we have received every possible
civility and attention, and most valuable assistance.

The Lyells joined us here on the 10th, and on the 1854. morning of the 12th, Charles Lyell sailed in a vessel hired for the purpose, for the island of Palma, where he was anxious to visit the great crater, which is very difficult of access. Mary remains with us.

We have paid three visits to the great Dragon tree, of which Humboldt gives such an interesting account, it is in a garden at the Villa de Oratava, about three miles from hence, and is a noble ruin of a tree. One half of it was blown down in 1819, and much has perished at various times since, but what remains is still very grand as well as interesting. At first sight one is more struck with the singularity of its appearance than with its size, but when one goes close up, it appears enormous. We measured it, and found its circumference, round the part of the trunk which remains entire, to be 30 feet, but so much is gone, that I should think, when entire, it must have been fully 50 feet round. This measurement was made at about 9 feet above the lowest part of the trunk that is visible above ground, for there is a terrace against one side of the tree. It is probably one of the oldest trees now existing in the world. Pray look at Humboldt's account of it, in the first volume of his Personal Narrative, chapter ii., and in the second volume of Aspects of Nature. There are many other Dragon trees in this part of the country, and some very fine ones, but comparatively young trees, vigorous and symmetrical. One that I measured yesterday is 14 feet 4 inches round, at 4 feet from the ground.

1854. They are strange stiff-looking trees; when quite young and before they begin to branch, they are very like Yuccas, but afterwards the swollen, gouty appearance of the branches, and the way in which these are arranged, radiating like the spokes of an umbrella, without anything like a leading shoot, give the tree a very peculiar character.

We have got some cones of the Canary Pine, which is a noble and most picturesque tree, one of the finest of the Pine tribe that I have ever seen. I am not sure whether it will bear the climate of Barton, but I hope that the seeds will vegetate, and that you will be enabled to try the experiment. It has become much less abundant in this island than formerly, but still exists on many parts of the mountains. We saw some fine trees of it in a very interesting excursion we made the other day to the spring of Agua Mansa, which is nearly 4,000 feet above the sea, in a beautiful wooded ravine of the mountains to the south-eastward of this place. I was charmed with the beauty of the tree heath, Erica arborea, which is the predominant plant in those woods, and is now in profuse blossom. It grows really to a tree, and the effect of its innumerable myriads of white bells is quite lovely. We are both very well. We hope to remain here another fortnight or thereabouts, then return to Santa Cruz, where Lyell will have many things to examine in geology, and to embark in the homeward-bound steamer which will touch there about the 4th of April. I have given up all expectation of reaching the summit of the Peak, but I hope to get as far

as the plain of pumice which surrounds the cone, at 1854.
the height of 8,000 or 9,000 feet above the sea.

Pray give my love to Emily, Cecilia, and Henry,
and believe me ever

Your affectionate Son,

C. J. F. BUNBURY.

JOURNAL.

Teneriffe,
February 18th, 1854.

Began to see the island in the forenoon, but
faintly; by-and-bye could distinguish the high north-
eastern extremity of it. The Peak entirely hidden
by clouds. The appearance of the northern part of
the island as we first approached it, that of a black
frowning mass of high and craggy mountains. A
low sloping point of land (Punta Hidalgo) seen in
the distance westward. Running along the N. E.
coast, saw well the succession of lofty abrupt headlands, running down from the inland mountains.
Striking variety of strongly marked colours in
the sea cliffs : — dark red, brown, blackish, and
whitish-grey. Anaga Point, Antequera Point, a
high craggy peninsula, connected with the main
land by a comparatively low neck : here, and still
more in the next, Roquet Point, remarkable display
of grand volcanic dykes, running up through the
cliffs to a great height, and exceedingly conspicuous.
Beautifully bright verdure in some of the deep and
narrow valleys opening to the sea between the
mountainous headlands, terraced cultivation ; village

1854. of S. Andres at the mouth of one of these valleys, with its white houses with red roofs. The coast mountains, though excessively steep, more verdant than I had expected to see; thickly dotted over with round pale-green bushes or trees (Euphorbia Canariensis, Mr. Hartung says), producing a singular spotty appearance. High dark serrated ridges, rising into wild crags and peaks in the back-ground. Town of Santa Cruz conspicuously white, standing on nearly flat ground apparently; batteries and numerous wind-mills: to the S. W. of the town, two or three insulated dome-shaped hills,—rounded or flattened cones, reminding one of the Picos near Funchal. The mountains to the W. of the town, comparatively low and tame; to the N. W. and N., a succession of dark threatening craggy ridges. Perilous landing at the quarantine station, S. of town, a heavy surf though the day was calm. Beach even more precipitously steep than at Funchal.

February 20th.

Behind the house in which we are performing quarantine (which is on the sea-shore close to the Lazaretto and the Fort) there is a small space of flat waste ground, partly stony and partly muddy, on which we are allowed to walk.

February 22nd.

Santa Cruz. We walked out along the coast road to the N. of the town, as far as the fort of Paso Alto; then turning to the left, ascended the

narrow valley, or torrent bed, which comes down 1854.
directly behind the fort, and which our guide called
the Barranca de Paso Alto; but it seems to be
the same which in Captain Vidal's map is called the
Val Ameida. The lower part of it is quite dry at
present: in the upper part there is a pretty little
clear stream which quite disappears before reaching
its entrance. But the huge rolled stones which
encumber the bed, show that the torrent must at
times be formidable. The first remarkable plant
which we saw on the rocks at the side of the
ravine was Plocamo pendula, which continues very
plentiful all the way up, scattered over the rocks
and hills, a plant of a very peculiar and well-marked
aspect which I recognized at once from descriptions. It forms bushes usually under 5 feet high,
but quite shrubby, excessively branched, with long
slender, pliant, gracefully pendulous branches, and
slender almost cylindrical leaves, the whole of a
lively grass-green colour, and having altogether the
look of a miniature weeping willow. The small
bell-shaped greenish white flowers at the ends of
the shoots, are now just opening. The rocky and
stony sides of the hills are thickly dotted over with
large clumps of the curious Euphorbia Canariensis,
which has exactly the look of those Euphorbias
that predominate so much in the "Bush" on the
Cape frontier. Its thick succulent, leafless, angular
stems, growing in very thick clumps, rise from the
ground at first in curves, and then become erect, all
rising nearly to the same height, or when growing
at the edge of a rock, they hang down some way,

1854. and then rise up with a curve, like the branches of a
chandelier They have either four or five angles,
nearly as often the one as the other : the angles a
little waved, beset with short prickles placed in
pairs, and they often run in a spiral direction,
the sides of the stem between them a little concave
and very smooth. The colour of the younger part
of the stem is bright-green, but they soon get a
greyish or whitish hue with age, so that the general
tint of these Euphorbia clumps is a rather pale
whitish-green, and their appearance, thickly sprin-
kled as they are over the hill sides, is most singular.
The milk gushes out in vast abundance on the
least incision. It is said to be excessively caustic,
but I did not try. The plant is called *Cardon* by
the people here. Together with this and the
Plocama, there grew on these hills abundance also
of the Kleinia nerrifolia (as I suppose) another
singular plant with a tall, thick, fleshy, smooth
gouty-looking stem, and long narrow glaucous leaves.
This like the Euphorbia, is at present quite out of
flower. Sometimes we saw both this and the
Plocama growing in close contact with the masses
of Euphorbia, and forming a curious combination ;
the graceful weeping form and lively green colour of
the Plocama, contrasting strongly with the grey hue
of the Kleinia, and with the stiff columns of the
Euphorbia. There was also abundance of a
beautiful cut-leaved Lavender, with deep violet blue
flowers,—I suppose Lavender abrotanoides ; a white
flowered Pyrethrum or Chrysanthemum in great
quantity ; a handsome purple-flowered Senecio (or

Cineraria), and many other things. Indeed the 1854.
number of plants in flower was quite remarkable.

We saw in this walk two or three long trains of
loaded Camels, a very Oriental or African sight.
Saw many Kestrels, both in the ravine and about
the roads nearer the town : surprisingly tame and
fearless. The basaltic dykes in the ravine of Paso
Alto are some of the most conspicuous and remarkable I have seen anywhere. We observed as many
as eight or nine of them. Three I examined and
measured which are crossed by the stream, and seen
running nearly vertically up through the rocks on
either bank. The first 9 feet 2 inches wide, of very
schistose but otherwise very compact basalt, of a
blue-grey colour, traversing red lava which is very
scoriaceous, so that the contrast both of colour and
texture is very striking; the schistose structure is
parallel to the walls of the dyke. The second 6
feet wide, of very compact and hard basalt, of a
distinctly prismatic structure, with its prisms transverse to its direction. The third, 4 feet 8 inches
wide, of similar compact blue-grey basalt, trans-
versely prismatic in the middle, and schistose
(parallel to the walls) at the sides, it traverses a red
scoriaceous breccia.

———

February 24th.

I visited M. Berthelot, and found him very lively
and conversible, talking very readily and fluently on
botanical subjects. His remarks on the botany of
the Canaries were very interesting. He particularly
recommended me to visit the wood of Agua Garcia,

near the road to Orotava, which he says is easy of access, very beautiful, and particularly rich in plants. The tree Heath grows there 45 feet high. He says there are not less than 25 species of forest trees here, and most of them peculiar to these islands. He showed me a fine collection of sections of Canarian woods. The inhabitants of the islands, he says, have distinctive names for all the more remarkable native plants, and these names often very expressive and appropriate, and even showing an acute perception of analogies.

Afterwards we rambled out to the coast south-ward of the town, first passing through a considerable extent of uninclosed cornfields, where the bearded Wheat was already in blossom. We botanized among the rugged, black lava rocks on the headlands of the coast, where the Euphorbia Canariensis grew in great abundance, its strange looking clumps of stiff, green, fluted columns springing up vigorously amidst the most barren rocks. But I did not see a trace either of the Plocama or the Kleinia, or the beautiful Lavender, which grew in company with the Euphorbia, in the valley of Paso Alto. The most interesting plant, not before seen that we met with here, was Matthiola parviflora, which grew scattered not unfrequently among the arid volcanic rocks, both in flower and fruit. Aizoon Canariense (if I am not mistaken), was plentiful among the rocks.

February 25th, 1854.

We went a good way along the coast to the north-eastward of Santa Cruz, passing successively the

valleys of Paso Alto, Valleseco, Bufadera, and 1854.
La Paja; climbing by very bad roads over the
lofty and precipitous headlands between them.
This coast scenery is wild, striking and grand. A
succession of mountain buttresses run down from
the great north-eastern dividing range, all ending
towards the sea in very bold and savage cliffs; or
perhaps the whole coast might be described as one
range of mountain cliffs, only interrupted where the
deep and narrow valleys or ravines open down to the
sea. The strangely rugged and fantastic forms and
gloomy blackness of the lava rocks; the strange
vegetation; the surf perpetually dashing against the
foot of the cliffs; the hawks circling around them;
all make it a striking scene.

The predominant plant all over these cliffs and
coast mountains, and that which gives them the
most strongly marked character, is the uncouth
Euphorbia Canariensis; certainly one of the strangest looking plants I ever saw. It is in prodigious
abundance here, and very vigorous in its growth,
and forms huge clumps. Here and there it was
coming into blossom : the flowers very small in
proportion to the plant, of a dark, dusky red colour,
are sessile on the angles of the stem.

Our guide or donkey-man spoke with great horror
of the venomous properties of the "*Cardon;*" said
that it was "muy malo," and that a drop of its milk
falling into the eye would produce blindness. A
huge black spider is very often to be seen amidst
the clumps of this Euphorbia, weaving its large and
very strong web from one stem to another. The

1854. graceful Plocama pendula is likewise very common
all along this coast, not only in company with the
Euphorbia on the cliffs, but also in the stony torrent
beds, where however it is often battered and mu-
tilated.   It is called by the country people *Balo.*
Its stems have a very offensive smell when broken,
reminding one of its botanical affinity to the
Coprosma.   I did not see much of the Kleinia to
day, but there was great plenty on the cliffs of a
large handsome, shrubby, glaucous Euphorbia,
which the people call Tabayva, or Tavayva; I
suppose Euphorbia piscatoria; it has a good deal
of the look of the Mediterranean Euphorbia dendroides, but the stem is thicker and more succulent.
The delicate glaucous tint of the leaves is pretty.   It
is almost as copiously milky as the Euphorbia
Canariensis, with which it grows in company.   We
noticed also great abundance of the beautiful Lavandula abrotanoides, both on the cliffs and down in
the stony beds of the torrents.   Also a strong
scented silvery-leaved wormwood, which our guide
called *Incienso,* and praised for its medicinal qualities; an infusion of it in hot water, used as a foot-
bath, was excellent he said for relieving headaches
and colds in the head.   Aizoon Canariense and
Plantago Amplexicaulis, frequent on the rocks.
We also met with a pretty little dwarf Cyperus,
which Fanny had found before on the 23rd ; and I
was surprised to see a plant of that genus thriving
on such arid rocks.   I remarked the absence not
only of Ferns, but of the two large Houseleeks
which are so very general on the rocks of Madeira.

Kestrels are very numerous along this line of cliffs 1854. and very fearless ; one perched to-day on a ledge of cliff within a stone's throw of me. I saw also a larger kind of hawk; and large birds of the crow kind (carrion crows (?) or ravens (?), are frequent. We discovered a remarkable deposit of fossil land shells (Helices) in the high promontory between the Val de Bufadero and the next valley to the N. E. They are imbedded in great numbers in a volcanic breccia composed of large and small cinders and angular pieces of scoriaceous lava and volcanic dust. Where the substance happens to be friable, the shells are easily got out entire, but the greater number are inextricably imbedded in the rock. We observed three apparently distinct kinds, but that which greatly predominates is a large strong Helix of a very round form. The numbers of this are very great ; the face of the cliff appears quite studded with them.

February 27th.

We rode from Santa Cruz to the City of Laguna, the ancient capital of the island ; distant (in a straight line), between five and six geographical miles in a N. W. direction. The road broad and good, partly paved, but well made and in good order. The ascent most part of the way quite gradual ; the only considerable hill is that called La Cuesta, where one first ascends from the maritime plain of Santa Cruz to the hills ; and even this is not steep. The scenery is not interesting ; country very bare ; a good deal of cultivation, but entirely

1854. without the rich luxuriance of Madeira ; fields of green corn, and of the Cactus (Cactus Tuna (?) or Ficus-indica (?) extensively cultivated for the sake of the cochineal, which is now the most valuable produce of the island. This is far from a beautiful or picturesque crop. Fields divided by low walls of black lava ; no trees or hedges. The hills too along this road are for the most part tame.

Laguna, a small town of a most melancholy and decayed aspect, situated in a plain, elevated 1726 feet above the sea (according to Captain Vidal's map) ; the air very perceptibly colder than at Santa Cruz. Yet there are some Palm trees about the town apparently flourishing though not tall ; and in a garden at the village of Las Mercedes, which is still higher, we saw Palms, Orange trees, and a Dragon tree.

The hills bounding the plateau on the E. and W. are steep, but of moderate height and very bare ; some of them flat-topped or table-shaped, others of very irregular shapes. To the N. the plain contracts into a narrow valley, running up past the village of Las Mercedes to the hills on which stands the wood of that name. To the S. the hills rise much higher, joining on to the great central chain of mountains ; and from Las Mercedes we *ought* to have seen the Peak, towering above all the other mountains in this direction ; but it was entirely hidden by clouds. Between these mountains to the S. and the hills to the W. of the plain, there appears to be a gap or outlet.

Luxuriant and striking vegetation on the walls

and on the roofs of the old houses at Laguna. A
Sowthistle of immense size (Sonchus congestus,
according to Webb and Berthelot), and a very fine
large Houseleek, I presume the Sonchus urbicum,
which was first discovered in this locality by
Christian Smith,—are the most conspicuous. We
saw also great plenty of Polypodium vulgare, the
first Fern we have seen in the island, except some
wretched starved morsels of Adiantum capillus-
veneris on a wall at Santa Cruz. The common
Groundsel also flourishes prodigiously on these
walls.

In the valley of Las Mercedes the Arundo Donax
cultivated, and a kind of Lupine, and the same
Caladium which is so much cultivated in Madeira
under the name of *Tuhanie*. It was too late in the
day for us to go on to the forest of Las Mercedes,
but we went far enough to see many scattered trees
of Laurus indica, the Vinhatico of Madeira,—relics
no doubt of its former extension. In this upper
part of the valley we observed abundance of a
beautiful crimson or purple flowered Cineraria (Cin.
tussilaginus (?) which our donkey-man knew by the
name of *tossilago* ; a nearly shrubby Andryala ;
Pteris aquilina ; Asplenium Adiantum nigrum ; and
on the steep shady bank of the little stream, a large
Fern with the habit of a Nephrodium, but without
fructification—apparently different from any of those
we collected in Madeira. In the plain, Oxalis
cernua with *double* flowers, was growing as if wild in
some of the fields.

Santa Cruz.

We made an excursion up the valley of Bufadero,
which opens on the coast some miles north-east
of Santa Cruz. In the lower part, the whole
breadth of the valley is occupied by a wide bed of
shingle, brought down at different times by the
torrent, which now disappears among the stones
before reaching the sea.

Some way up, we come to the point of junction of
two ravines, which unite to form the valley. Just
at the junction there are some gardens full of beautiful Orange trees, the blossoms of which gave out a
delicious smell. We took the left or northern
branch, and found it a very picturesque ravine,
narrow and winding, with bold, steep rocky sides, a
beautiful little clear stream dashing over rocky
ledges, a variety of curious shrubs and pretty flowers
adorning the margin of the rivulet and clothing the
rocks on each side, and the hills rising into strange
wild crags and peaks above.

The Tabayva is remarkably abundant, and of
very vigorous growth in this ravine It is a very
handsome shrub, or indeed small tree; the trunk,
covered with a smooth grey bark, rises singly and
erect from the root to the height of 2 or 3 feet, then
divides into numerous branches, all springing from
one point, and again and again divided; the
ultimate branchlets alone being crowned with tufts
of delicately glaucous leaves. The general form of
the whole plant is almost that of an inverted cone.
It here grew 10 or 12 feet high, with a trunk 2 feet

in circumference. The Euphorbia Canariensis also 1854. was here in prodigious abundance ; its clumps often beautifully wreathed by the twining stems of an Apocyneous or Asclepiadeous plant, — I believe Periploca laevigata,—with leaves like those of an Echites, and curious long horn-like follicles. On the rocks grew three or four different kinds of Sempervivum, one of which, a small delicate species, was very pretty with its starry golden flowers ; the same beautiful Cineraria which we had found at Las Mercedes : the handsome Lavandula pinnata, which we found in Madeira ; the Lycopodium denticulatum and in one place the Adiantum capillus-veneris, Kleinia neriifolia here and there, and various other shrubs, which, without flowers, I did not recognize. The vegetation grew more and more rich and various as we ascended further. At last we came to a beautiful spot, where high up the narrow ravine, were several good houses and pretty gardens ; superb Orange trees loaded with fruit and flowers, Lemon trees, Custard Apples, Bananas, and two fine Palm trees : quite a Madeira scene, with the wild craggy hills towering above, and the pretty stream overhung by a variety of native shrubs, dashing from rock to rock below. At intervals all up the valley, the Cactus is cultivated, often ex-tending a good way up the hills, and mingling with the native Euphorbias. But I have not yet, either here or in Madeira, seen Cactuses as large as some I saw in Sicily.

We saw some partridges (red-legged, I believe) and large lizards ; and heard much croaking of frogs,

1854. which I should not have expected in such a situation.

_____

We revisited the Val Bufadera, exploring it a good way further up than last time. It continues very interesting; beautiful spots of cultivation, rich in Orange and Lemon and Peach trees, Bananas and the tall Reed,—alternating with wild rocky scenes. The native vegetation too is very interesting; the mixture of Kleinias and Euphorbias, Ferns, House-leeks, Echiums, Lavenders, and strange exotic composites clothing the rocks.

The basaltic dykes in the Val Bufadera are innumerable, and many of them extremely remarkable standing up so high above the surface as to be conspicuous from far off, looking like ruined walls, and running up in some cases from the bottom of the valley to the very tops of the hills. Sometimes, by the denudation of the surrounding rock, the vertical side of a dyke is exposed for some distance, rising from the valley in a smooth bare sheet of rock.

On the summits too, they stand out in most extraordinary fantastic crags. But they are not always vertical. I observed several instances of the smaller class of dykes, running obliquely or almost horizontally, and even in curved and waving lines, through the tufas and scoriaceous aggregates. The structure is most often prismatic, but sometimes remarkably schistose.

_____

March 6th.

From Santa Cruz to Puerto de la Orotava ; a journey which took us 9½ hours, including about an hour for rest and refreshment. The distance is called seven leagues, which should be only twenty-one miles, but it is probably not less than twenty-four or twenty-five miles. We went in a carriage as far as the road is practicable, which is about two miles beyond Laguna ; afterwards, I on Mr. Murray's pony, Fanny sometimes on a donkey, sometimes in Mr. Smith's hammock. Almost as soon as we had passed Laguna, the Peak came full into view, above the bare green hills on our left; a smooth symmetrical cone, covered with a sheet of brilliant snow.

A little further on our way we saw it still better, covered with snow on the north side at least, quite down to the base of the steep cone; towards the south, the snow, at least the continuous unbroken snow, did not seem to reach quite so far down. The numerous ravines and furrows of its sides very conspicuous through the snow.

From Laguna to near Tacoronte. The road runs through an elevated plain, all cultivated, but very bare and open, without trees or enclosures ; fields of Wheat and Lupines. The country would be very uninteresting but for the view of the Peak. Presently the sea appeared on our right ; afterwards we descended a little into a more enclosed and cheerful country.

Here we saw Asplenium palmatum growing on the

1854. stone walls, rather plentifully, but of smaller size than on the north side of Madeira.

Village of Tacoronte below us on the right, and further on Sauzal, in a similar situation. The country descending with a moderate slope from the mountains on our left to the sea, richly cultivated many Palm trees.

After passing Sauzal the view of the Peak was glorious; the long ridge below half lost in the bright hot haze, and light clouds floating about its middle; while the cone with its dazzling snow, seemed to soar to an immeasurable height into the sky.

About and beyond Matauza, the country very rich and beautiful, and becomes more and more so as we proceed westward by the villages of Vitoria and Sta. Ursula, towards Orotava.

Between Vitoria and Sta. Ursula it is particularly beautiful. Abundance of Date Palms, Orange and Fig trees, Peach and Almond and Pear trees in blossom, Bananas and tall Reeds, and bright-green fields of Wheat and Flax: high wooded mountains towering above us on the left, and frequent deep rocky ravines running down from them and inter-secting the garden-like cultivated lands. This part of the country reminded me much of Funchal. But Palms are far more numerous here than in Madeira.

A little beyond Sta. Ursula, from the brow of a hill a glorious view opened upon us; the broad beautifully rich and green garden-like valley of Orotava, with the glittering white houses of the Villa de Orotava in the midst: the valley sloping up

from the sea to the great wall of dark and lofty 1854.
mountains behind : and the glittering snow of the
Peak towering high over all.

Puerto de Orotava.

Mr. Charles Smith took us to see the Botanic
Garden, which is a little way above and to the N. E.
of the town, in a fine situation. It is in a sadly,
neglected, forlorn, ruinous condition, with a few fine
trees, and here and there some handsome plants
running almost wild amidst the general disorder.

In particular a fine tree of the Pinus Canariensis,
a very handsome Pine, with a cinnamon-coloured
bark, a bold and free growth, large cones and very
long slender, drooping, light green leaves. Also
another Pine, name and country unknown, and
which I did not recognize ; a tree of very peculiar
growth, with a comparatively low trunk, and very
wide-spreading, drooping, almost weeping branches,
forming a remarkably broad umbrella-like head.

In Mr. Smith's garden, and also outside the
Botanic Garden, we saw many Dragon trees of
various ages, some of considerable size and much
branched.

Most strange, uncouth trees they are ; I always
fancy them fit to have been contemporaries of the
Iguanodon and Megalosaurus. The mode of ramification is almost always *radiating*, a number of
equal branches springing from one point, without
any central or leading shoot ; and this mode of
division is repeated again and again, but as the

1854. plant grows older, the branches become more and more irregular, swollen and distorted, sometimes swelling towards the top, sometimes in the middle, sometimes swollen and pinched in alternately. The young plant, before it branches, has very much the look of a Yucca.

March 8th.

We rode with Mr. Charles Smith to Villa de Orotava, not by the main road, but by a very bad stony road, in part indeed along the dry, stony bed of a torrent. Passed close by an insulated volcanic hill, called the Montanēta de la Villa ("Cinder Hill" in Captain Vidal's map) a large somewhat saddle-shaped mound of black cinders, almost bare of vegetation, exceedingly arid and black and savage looking. Another volcanic mound at some distance on our right, the Montanēta de Los Frayles.

The Villa de Orotava, a handsome but rather decayed looking town, stands about one thousand feet above the sea. We visited the famous Dragon tree, of which Humboldt has given so interesting an account. It has been exceedingly shattered, and has lost more than half its top since his time, but is a noble ruin; the trunk now a mere shell built up with masonry in the interior, to support it; but the foliage of the top still appears fresh and vigorous. I looked with extreme interest on this venerable tree, not only on account of its immense age and great celebrity, but because of the delight with which I had read Humboldt's account of it. Though infinitely more picturesque than any other Dragon tree I have ever seen, it still retains something of the usual uncouthness and grotesqueness of the species, in the swollen, distorted, gouty appearance of the ultimate branches, and the stiff tufts of leaves at their ends. The lower part of the trunk is exceedingly rugged, the branches all rise up at a high angle, so that the general form is nearly that of half a cone inverted, and the top is not far from flat. The divisions of the older branches are very irregular, but towards the top the dichotomous ramification becomes evident. The garden in which this famous tree stands, belongs to the Marquessa del Sauzal, widow of the late Marquis, whose death has caused great affliction here. He was of the family of Cologan, the son, as I understand of Humboldt's friend, and inherited also the estates of the family of Franqui.

In the same garden is a remarkably tall and beautiful Palm tree; I am not sure of the species; the form of the leaf-scars, and appearance of the stem are like those of the Date Palm; but if it be of that kind, it is much the tallest I have ever seen. It has a singular contraction towards the middle of the stem.

From the Plaza, in the upper part of the town, a beautiful view over the green, fruitful slopes of the valley, the houses and gardens of the town, and the blue sea.

From the Villa we rode on westward, through a lovely, rich, smiling country, to the village or small town of Realejo. The walls of loose stones by the

1854. wayside are beautifully clothed with Ferns and House-leeks. Davallia Canariensis and Polypodium vulgare in great profusion, very luxuriant and beautiful; the delicate little Gymnogramme leptophylla, very plentiful, forming conspicuous tufts of a peculiar tender pale yellow green; Asplenium palmatum also abundant; Notholæna Marantæ very fine. We found also Nephrodium elongatum. In one of the "barrancos" which we crossed I found Hypericum grandifolium in flower, for the first time; it is an abundant plant in this part of the country, as well as in many parts between Laguna and Sta. Ursula. The beautiful crimson Cineraria is likewise very plentiful and ornamental here. Realejo de Arriba is finely situated, near the edge of the rich valley, and almost overhung by the frowning steeps of the Tigayga mountains. Its elevation is probably nearly equal to that of Villa de Orotava. Its Church, said to be the oldest in the island, and built at the time of the conquest, in the 15th century. A very tall and beautiful Palm in a garden here. Hence we descend through Realejo de Abajo, where we noticed a very fine Dragon tree, and so back to the Port.

This was a lovely day, less hot than the two preceding, with a refreshing N. E. breeze. The Peak brilliantly conspicuous when we first started on our ride, but soon after noon the clouds collected round and quite hid it.

———

March 9th.

We rode along the coast westward as far as the

Playa de Callado. The road was at first the same 1854.
as that by which we had returned yesterday, nearly
as far as Realejo de Abajo. Cross several of those
deep rugged *barrancos*, or dry torrent beds, which
are so characteristic of this country : first near the
town, the Barranco de Cabezas, next the Barranco
del Patronato, then the Barranco de Realejo. In
the second of these, we found a beautiful little
Cheilanthes growing among the rocks, much larger
than the Madeira Cheilanthes, yet possibly not a
different species. It is certainly entitled to the
name of *fragrans*, for after being some time in the
tin box, it had a strong fragrance of hay.

Hacienda de Castro, low down near the sea, a
whole grove of Palm trees, and a remarkably
beautiful Palm standing by itself near the houses,
with its large golden bunches of fruit ; also a very
fine tall Dragon tree, remarkably symmetrical in its
form, with its numerous branches radiating from one
point, and forming a dense nearly level crown. Not
far off, higher up on the side of the hill, is
another Dragon tree, of quite a different aspect,
exceedingly irregular, contorted and picturesque.

The Kleinia neriifolia is very abundant on this
part of the coast, and in some places the Euphorbia
piscatoria, but not the Euphorbia Canariensis nor
the Plocama.

Beyond this we descended by a steep road to the
seashore, and continued for some way along the
beach, skirting the bold cliffs in which the long
mountain ridges that run down from the great
central mass, terminate towards the sea.

A cool cloudy day. Revisited the Villa de
Orotava, and the great Dragon tree, of which Fanny
made a sketch. With the assistance of the
major-domo we measured it, and found its circumference to be 42 feet, at about 8½ feet from the base;
but about 12 feet of this was measured in a straight
line across the hollow of the trunk, and therefore
was only the chord of the arc. The circumference
when the stem was entire must I should think, have
been at least 50 feet. There is a terrace against
one side of the trunk, so that the height measured
from the ground is different on the two sides; the
8½ feet I have mentioned were measured from the
base of the trunk when the ground is lowest.—
Numerous tufts of air-roots are protruded from the
branches, and even here and there from the stem,
but they do not grow to any considerable length.
Many of the branches are singularly tortuous, and
some which run to a great length without dividing,
are so remarkably contracted at intervals, as to
suggest the very unpoetical comparison of a string of
sausages.

We visited the beautiful spring of Agua Mansa,
situated in a ravine of the great screen of mountains
at the head of the valley of Orotava, nearly south-
east of the Villa. We started from the Port about
half-past eight a.m., reached the waterfall about
noon, spent about two hours in that neighbourhood
and got back to our hotel about half-past four p.m.

The morning was splendidly bright and clear, and 1854. the snow of the Peak not obscured by a single cloud, shone out gloriously against the blue sky ; but we had scarcely been out an hour before the clouds began to gather round it, and it was soon totally concealed.

After passing through the Villa de Orotava, we we had a long ascent by a very rugged, rocky and stony path (apparently the surface of a lava current). to the top of the first zone of hills behind the town. These are almost everywhere cultivated, and towards their tops are abundance of scattered Chesnut trees.

At these elevations, the trees, both Chesnuts and Fruit trees, are at present leafless, and what with this and the quantity of brown soil, fallow or newly broken up, and the prevalence of black lava walls between the fields, the character of this part of the scenery is rather sombre. Hypericum grandifolium is here one of the most common plants, and on the stone walls, particularly in the lower part of the ascent, there is great abundance of the beautiful Davallia, and of the Polypodium vulgare in great luxuriance. Higher up, scattered plants of Erica arborea begin to make their appearance.

At length, after ascending a long way, we quitted the enclosures, and came on to a sort of plateau, covered with a most beautiful luxuriant shrubbery of Erica arborea in full blossom. This heath was ten or twelve feet high, or more, and the effect of its countless myriads of white bells was perfectly

1854. charming. Tall bushes of Myrica Faya were intermixed with it.

Before us rose the precipitous face of a mountain, furrowed in a very extraordinary manner, with deep, dark, strictly vertical chasms : hollows left by the removal of dykes (?). They seemed too perpendicular for mere water-courses. Yet, as far as I have seen in this country, the volcanic dykes are always harder and more durable than the rocks they traverse.

We now passed a farm house and some very large Chesnut trees, and entering the ravine of Agua Mansa, ascended through a beautiful wood to the head of the water-course. This wood which is very thick and luxuriant, consists in great part of Erica arborea of really tree-like dimensions, and most beautiful with its profusion of blossoms. Intermixed with this is great abundance of the Faya, a very handsome evergreen shrub or small tree : also the Açevinō, a species of Holly,—a Viburnum nearly allied to the Laurustinus (V. rugosum) and some large trees of the Vinatico Laurus indica. Hypericum grandifolium extremely abundant, and one or more species of Cistus, not now in flower. I saw no trace of the Madeira Vaccinium. The ground, the rocks, and the trunks of trees are richly carpeted with mosses, showing the moisture of the climate, and the branches are hung with long streamers of Usnea. The beautifully clear stream forms pretty cascades over the rocks : and from it by means of acqueducts, the Villa and the whole valley of Orotava are supplied with water. The elevation of Agua Mansa, according to Webb and 1854. Berthelot, is 3820 feet, I suppose French feet. At this height the air was very chilly, and we were repeatedly enveloped in clouds, and exposed to a drizzling rain.

On the margin of the stream grows a beautiful Forget-me-not, very like our Myosotis sylvatica. A handsome Cineraria, I believe the same that is common in English greenhouses, with leaves of a fine purple colour at the back, is abundant in the wood, but not yet in flower.

On the more open slopes of the mountain, adjacent to the ravine of Agua Mansa, there are many fine Pine trees, Pinus Canariensis; the first I have seen in a wild state. We collected cones and young branches, and fragments of the wood, which is of a reddish-yellow colour, and extremely resinous. This is a remarkably noble and picturesque Pine, one of the finest I have seen. Its habit is quite different from that either of the Pinaster or the Aleppo Pine. The cones are in shape and size much like those of the Pinaster, but without the pointed knobs. Not far from these Pines, we observed a beautiful flowering tree, which the people of the country call *Escobon;* it is a Cytisus, much about the size of the Laburnum, with graceful weeping branches, and a profusion of beautiful white flowers. The vertically furrowed precipice before mentioned is part of the mountain of Perexil, or Pedro Gil, as it is variously called; the ravine of Agua Mansa is a little to the left (as we look up) that is, to the north-east of it.

March 12th.

Third visit to the great Dragon Tree. Mary made
a sketch of it, and Fanny went on with her drawing.
The bark of the trunk scales off rapidly, in very
irregular pieces, and is entirely bare of cryptogamous
plants.   On the smooth surface of the younger
branches one can distinguish from below, the
annular scars (appearing like slight wrinkles) caused
by the fall of the leaves.   The peculiarly tortuous
growth of many of the branches appears to me to be
owing to the developement in each case of a bud
which is not strictly terminal; whence instead of a
di- or trichotomous ramification taking place, as
would be normal, the axis is continued by a single
branch, which is not in a straight line with the
previous growth.   The morning and earlier part of
the afternoon were rather cloudy;   but towards
evening it cleared and the great belt of mountains
at the head of the valley of the Orotava had a
beautiful appearance, the sun lighting up their
crests, while the clouds formed a zone along their
flanks.   The Peak rose majestically above the
clouds that lay thickly along the intermediate
ridge.

March 13th.

We revisited in company with Mr. Smith the
two Realejos.   Realejo de Abajo is a beautiful spot.
Gardens rich with the beautiful foliage of Bananas
and Orange and Coffee trees, and Palms, and huge
fig trees, rise in steps one above another up a steep
hill, crowned with white houses and convents and

churches ; and high above them towers the gigantic 1854.
wall of the Tigayga mountain.   We entered a large
garden, and rested under a thick shade of mingled
Orange and Lemon trees, loaded alike with flowers
and fruit.   There were in this garden, also abundance of flourishing Coffee bushes in full fruit—but
not of finer growth than at Madeira.

In the lower part of Realejo de Arriba (not
in that de Abajo, as I wrote by mistake before),
by the side of a convent, stands a very fine Dragon
tree, in full vigour and freshness of growth with a
very full, close regular head.

I measured the trunk and found its circumference
to be 14 feet 4 inches, at 4 feet from the ground.

This was a lovely day ; the temperature and feel
of the air were quite delicious.   At night the view
of the Peak by moonlight, from the terrace roof or
Azotea of the hotel, was exquisitely beautiful, as its
brilliantly white pyramid was seen relieved against
the clear dark sky, while the long wall of lower
mountains above which it towered was in deep
shadow.

At Realejo de Abajo (the elevation of which is, no
doubt, below 1000 feet) we found Adiantum capillus-
Veneris in great beauty, and Asplenium palmatum.
Humboldt was certainly misinformed in restricting
the list of Ferns found in the region of the Vine in
Teneriffe, to two species of Acrostichum and an
Ophioglossum.

The Davallia, the Polypodium, the Gymnogramme leptophylla, and the Adiantum, are
plentiful even below the level of the Villa de Orotava,

3 T 2

1854. whereas the upper limit of the region of Vines must certainly be near 1000 feet above that town.

I have nowhere seen birds of the hawk kind so numerous or so tame as in this island. Kites and Kestrels are especially common, and this morning a large Kite alighted in a *barranco* within a few yards of us.

The Castor Oil tree, Ricinus, grows very luxuriantly here, and is apparently wild in some of the ravines; if not indigenous, it appears to be quite naturalized.

March 17th.

Weather very hazy and cloudy, and the clouds very low on the mountains, as has been the case the last four days. I have not seen the Peak since the 13th.

Ascended the two insulated dome-shaped volcanic hills, called the Montañeta de la Villa, or de Llarena, and Montañeta de los Frayles. The first, situated nearly due South of the Puerto, at some distance W. of the main road from thence to the Villa, is a double hill, with one of its domes or rounded cones much higher than the other, and a deep saddle-like depression between them. It is very steep and bare, and entirely composed (as far as one can see externally) of loose or very slightly aggregated cinders, excessively vesicular and scoriaceous, generally black and very light, quite spongy. Many of them have a rich metallic tarnish, blue, green, and purple, like certain iron ores. Others of the cinders are of a red-brown colour, heavier and rather less vesicular; and intermixed with them are some pieces of thoroughly ropy lava.

The Montañeta de los Frayles is situated about W.S.W. of this, and rises from a much higher base, as the whole country rises inland with a considerable slope from the sea. It is cultivated nearly to the top, so that its formation is not so apparent as that of the other, but it appears to be composed of loose cinders, rather less light and spongy than in the other hill. At the top it has a remarkable hollow, apparently the old crater; the rim complete all round, but considerably lower on the E. side. At the top of this hill I found a beautiful yellow-flowered Broom, with very small and numerous leaves, quite new to me; probably an Adenocarpus, for the pods are covered with remarkable glands. The Davallia and Gymnogramme leptophylla, very plentiful on the low stone walls on the sides of the hill.

From this height, one has a good view of the whole valley of Orotava. On the whole there is a want of wood, and the walls of lava by which the fields are divided, predominate too much in the landscape. I have nowhere seen here either the sugar-cane or the sweet potato, with which we were so familiar in Madeira. The principal plants culti-vated in the fields are, first, the cactus (for cochineal) and wheat; next, maize (which is now showing its beautiful broad leaves), beans, kidney beans, and lupines.

———

From Puerto de Orotava to Icod de los Viños: distance called 3½ leagues. As far as Playa de Callado, the way is the same I have already described.

At the further (Western) end of the beach called Playa de Callado, there is a large rock on the sea-shore, of remarkably columnar basalt; the prismatic structure more distinct than I have seen anywhere else, either in this island or Madeira. The columns are, as usual, of five or six sides; their direction oblique, approaching to vertical. On one side of the rock, where their ends are seen, the appearance presented by the junction of their polygonal figures is particularly remarkable. They are not separable, nor divided by regular transverse joints, as in the Giants' Causeway.

Among the rolled stones at the mouth of the Barranco Ruiz, between the Playa de Callado and S. Juan de la Rambla, Fanny discovered the Argemone Mexicana, said to be a very local plant in the Canaries. I had not seen it wild since Rio de Janeiro. Euphorbia Canariensis on the steep lofty rocks immediately above the little town of S. Juan de la Rambla.

From thence most part of the way to Icod a wild rugged, rocky and stony country,—a wilderness of stones: in appearance on a general view, very barren; but there is much cultivation, half concealed among the rocks and stone walls. In this rocky tract of country, we saw abundance of a beautiful white-flowered Cistus (C. Monspeliensis) and of the same beautiful Adenocarpus (foliolosus) which I found on the Frayles hill, also the lovely little Helianthemum guttatum, and an Acanthaceous shrub, a Justicia (?) with oblong, smooth, shining, bright green, rather fleshy leaves, and dull yellowish-white flowers. The curious Kleinia, and Euphorbia piscatoria in great abundance: but I saw no Euphorbia Canariensis. Adiantum reniforme (which we had not hitherto seen in Teneriffe), grew here and there on dry walls; and I found a beautiful Cheilanthes (?)—at least the habit is quite that of Cheilanthes, but the fructification rather of Pteris. Our *arrieros* brought me that very curious plant, Cytinus Hypocistis. We halted for a few minutes under a very fine and large Pine tree (Pinus Canariensis); on the hills near were many other scattered Pines, of less size: and at no very great distance above, we could discern the skirts of the great Pine forest, which I believe is the largest now remaining in the island. The general form and look of these trees reminded us of the Cembra Pines on the Wengern Alp.

Icod de los Viños, a pretty little town, situated in a beautiful fertile valley sloping down to the sea. Like other towns in Teneriffe, it is conspicuous from afar by its white houses. The streets wide and straight, very much grass-grown, with good foot pavements, and several large and handsome houses. The Tobacco plant grows apparently wild on the roofs of some of the houses.

Icod de los Viños.

A bright morning, glorious view of the Peak, much nearer and a far grander object than from Orotava. The truncated summit distinguishable. The lateral volcano, Chahorra, forming a projecting shoulder to the west, entirely covered with snow, as well as the Peak itself. Very soon the clouds came round the Peak, and we saw no more of it this day. The valley and nearest hills beautifully rich and tropical in appearance; numerous and tall Palm trees; the Anona and Rose Apple in the gardens, together with abundance of Orange trees and Bananas, several large Dragon trees. Below by the sea, a wide waste savage tract of bare and rugged lava.

We rode from Icod to Garachico, three or four miles further W. A long and very steep descent at first. Here I saw but in one place only, the Plocama which appears not to be a common plant on this side of the island. Near among the rocks grew a pretty Sisymbrium (?) with very finely cut hoary leaves and lemon-yellow flowers, and a delicate Paronychia with loosely-panicled flowers.

In one spot where a small rivulet crossed the road, Fanny discovered Pteris longifolia, a very fine and interesting Fern. We saw it nowhere else. Pteris arguta and Asplenium palmatum were growing in the same spot.

Asplenium palmatum and Adiantum capillus-Veneris descend to the level of the sea; the latter in great plenty and little before we come to Garachico.

I was struck with the rich and beautiful vegetation 1854. of the almost precipitous mountain that towered above us on the left, just before we reached Garachico, and especially with the sight of Bananas growing high up among the cliffs, in places seemingly inaccessible.

Garachico, a small town standing close to the sea, between two pretty little coves; one long and rather handsome street; some very large convents. It was formerly, it is said, a much more flourishing place, and had much the best harbour in the island; but the town was destroyed, and the port filled up by an eruption of lava in 1705 (or '6?), so that the present town, it is said, stands on the site of the former harbour.

The current of lava which occasioned this catastrophe is still very conspicuous in the landscape, still utterly bare, and rugged and desolate, threatening and awful, like a huge cataract of stones descending the steep mountain side.

I examined this lava at the point on the W. side of the town, on the sea-shore. It is like other lava currents that I have seen—excessively rugged on the surface, running into all manner of strange irregular fantastic projections, the external surface brownish, the interior deep black, rather shining on the fracture, hard and heavy, moderately porous, not containing any conspicuous crystals.

Argemone Mexicana abundant by the sea-side at Garachico, both on the lava and on the shingle of the beach; its bright yellow flowers have a gay appearance. A great profusion of the beautiful

1854. Statice and Frankenia already noticed at Playa de Callado.

We ascended the hill immediately beyond Garachico, and stopped at S. Nicholas, from whence we had a fine view of the coast to the westward.

Monte Taco, a large, solitary, very regular volcanic cone, strikingly similar to Monte Nuovo, near Naples, is a remarkable object in the view; the long low point of Buenavista, the most western headland of the north coast of the island, running out to seaward from it.

Date palms in fruit at S. Nicholas, the rich golden hue of the large bunches of fruit very ornamental.

March 20th.

We returned from Icod to Orotava by the upper road, passing through Icod el Alto. Weather very hazy and unfavourable; no view of the Peak. The road in many parts very rugged and difficult. The elevation of Icod el Alto, according to Capt Vidal's map, is about 1700 ft., and I do not think any part of our road was much above it.

In the wild, rocky and stony country, of which we traversed a great extent, the white-flowered Cistus and the Adenocarpus were in great abundance and beauty. The Helianthemum guttatum pretty plentiful. On the roots of the Cistus I saw plenty of the curious Cytinus, for the most part in rather an early stage, its yellow flower-buds just peeping above the soil, and still nearly enveloped in their bright red bracts.

Asplenium Adiantum-nigrum extremely common 1854. on the rocks and stone walls in this high country.

Between the village of La Guancha and that of Icod el Alto, we crossed a prodigiously deep ravine, the Barranco Hondo, the largest I have yet seen in this island. It is said to be the same which, towards its mouth, is known by the name of Barranco Ruiz. This chasm, with its lofty, precipitous walls of rock, richly mantled with shrubs and ferns, and a variety of strange and luxuriant vegetation, is both grand and beautiful. A little further on is another barranco, or a branch of the same (?) almost equally fine. Here the Tree Heath was in great luxuriance and beauty, in profuse blossom; and there were fine Laurel trees, Laurus Canariensis (?) and abundance of the Laurustinus which we had observed at Agua Mansa, very conspicuous and ornamental. Another great barranco, between Icod el Alto and the Tigayga mountain. These enormous ravines, with nearly perpendicular sides, are very striking features in the physical geography of Teneriffe. The bottom of the ravine generally forms a succession of steps, at different levels, the general slope of the waterway being interrupted from distance to distance by abrupt precipices of greater or less height, which must produce cascades whenever there happens to be much water in the barranco. These "saltos" appear to be occasioned by the intervention of a bed of very hard rock between softer strata. At this time there was little or no water in any of the barrancos we crossed, and they are said to be dry for much the greater part of the year; yet the

1854. smoothed and water-worn appearance of the rocky
channel sufficiently shows that there must be, at
times, a powerful action of water.

Descend the Tigayga mountain by a zig-zag paved
road of extreme steepness, and return by Realejo to
the Puerto. The day's journey was of about seven
hours, including a halt of some length.

March 22nd.

We rode out along the coast a little way to the
East of Puerto de Orotava, passing the villa of the
Cologan family, called La Paz, and examined a
ravine called (as I understand) the Barranco de
Llarena, which is curious in its geological structure.
It has, in the part we examined, two of those *saltos*,
or precipitous steps, which I have mentioned. The
precipice of the lower step is composed partly of
volcanic sand and tufa, more or less stratified, partly
of a conglomerate of large *rounded* (apparently water-
worn) masses of various volcanic rocks, in a sandy
basis; but at the top, overlying all these, is a thin
bed of very hard compact basalt, which, as I take it,
has preserved the softer and less coherent rocks
beneath it from erosion, and has thus been the cause
of the *salto*. Above this is a very thick bed—I should
think 40 or 50 ft. thick of perfectly columnar basalt,
of which the upper step consists, and which forms
the banks of the ravine from the one step to the other.

The columns are very regular, and of large size;
I measured a fallen one, of which the two largest
sides were each seventeen inches broad, and their
sides are as smooth and even, and the angles as
neat as if they had been wrought by art. They are

separable from one another, and are divided also
by transverse joints, though not as regular as in the
basalts of Antrim and Staffa; hence they readily
fall down, and where the rock overhangs, it is
curious to see some of these huge segments of
columns suspended as if about to fall, preserving
their cohesion with the rest only by a very small
part of their surface. The basalt is of a blue grey
colour, and very compact.

In the bed of the ravine (which is at present
quite dry) and also in the conglomerate, I found a
great number of pieces of a peculiar lava, containing a very great quantity of augite, in remarkably
large crystals, of a black colour and strong lustre,
together with much olivine in rather large yellow
grains. Some pieces of this lava contain
also long, narrow, white crystals, apparently of
felspar, and a small quantity of mica; others are
somewhat amygdaloidal and contain mesotype (?).

On the coast immediately to the east of the
Puerto, about La Paz and the Botanic Garden,
the basaltic lava which forms the sea cliffs is covered
by a light fawn-coloured earthy tufa, containing a
very great quantity of small pieces of pumice. This
peculiar tufa is largely developed also to the west of
the town, along the Realejo road, and between it
and the sea.

––––––

March 23rd.

In the Barranco de S. Felipe, which runs down
past the Montañeta de la Villa, I observed in a
large bed of basalt which borders one side of the

1854. ravine for a long way, appearances similar to those
we had repeatedly observed in Madeira. The
basalt is very compact, but the bottom part of the
bed is rough, scoriaceous, and appears to be composed of irregular fragments slightly cohering; and
the tufa on which it rests is of a brick red colour,
most vivid immediately under the basalt.

We found in this barranco, the beautiful white-
flowered Cytisus, " Escobon," which we saw at
Agua Mansa ; but here it was in a shrubby form, not
a tree. Erica arborea was also in blossom here, but
the only novelty we found was Lathyrus aphaca.

<hr/>

March 27th.

We made an excursion to the great plateau of the
Canadas, which surrounds the Peak at the elevation
of eight or nine thousand feet. The party—Mary,
Fanny, Miss Galwey, Mr. Charles Smith, and I,
with several arrieros and guides. Owing to a mistake about the horses, and some misunderstanding
as to the arrangements altogether, we started much
too late, not till very near 9 a.m.

The morning was then beautiful, and the Peak
perfectly free from clouds, looked splendid ; but
before ten, the clouds gathered round it.

We passed very near the Montaneta de los Frayles
and crossed at the village of La Cruz Santa, the
road leading from the Villa to Realejo de Arriba.
From thence a very long and steep ascent by an
excessively rugged, stony path, through the cultivated region of the mountains. Here, as well as in

the ascent to Agua Mansa, I observe that this first
stage above the valley, from about the level of the
Villa to the uppermost limit of cultivation, is one of
the steepest parts of the ascent. Towards the
upper limit the Tree Heath begins to clothe all the
waste and stony grounds which intervene between
the cultivated patches.

The plants in culture at the highest level are
Corn and Lupines ; but the Corn, which in the
valley is in full ear and approaching to ripeness, is
here but just above the ground.

The cultivated region is succeeded by an extensive
wild tract, beautifully clothed with the Erica
arborea (in full blossom), the Faya, and Hypericum
grandifolium, intermixed with Pteris Aquilina.
Both the Heath and Faya, however, are here of
dwarf growth, compared with what we saw at
Agua Mansa ; mere shrubs, not at all like trees.
I noticed also here and there the Acevino, a kind of
Holly growing as a moderate sized bush ; at Agua
Mansa it is a large tree.

This stage of the mountains has a much more
easy slope than the previous one ; for a considerable
space it may almost be called, comparatively at
least, a table-land. Our path constantly skirted a
very rugged and wild ravine, the Barranco de las
Rayas, (such I think was the name Mr. Smith gave
us for it) in which the ledges and beds of basaltic
lava presented very strange and striking appearances. A solitary Pine, high up in this barranco
was the only tree we saw after leaving the Fig trees
of the lower cultivated grounds. Formerly, how-

1854. ever, and even within the last half century, Mr.
Smith says, all the middle region of the mountains
was well-wooded.   As we ascended, the Heath
bushes became more stunted and scattered, and
after a time we perceived amidst them the first
plants of Adenocarpus frankenioides, one of the
most characteristic plants of the upper parts of these
mountains.   This was between twelve and one
o'clock, and the clouds, which for some time had
been gathering round the mountains now completely
enveloped us.

The Adenocarpus rapidly became more abundant
as we ascended, (for the ascent though steeper in
some parts and more gentle in others was continual)
for some little space it was intermixed with the
Heath, but this latter soon thinned out, and dis-
appeared, and for a very long space, certainly I may
say for miles before we reached the *Cumbre*, (or
ridge) the Adenocarpus prevailed to the exclusion of
everything else, covering the ground like Furze on
some of the commons and waste lands in England.
Its multitudes of short lateral branches, and the
very minute greyish-green leaves which cover the
branches as closely as possible, give it a peculiar
appearance; but although it has no thorns, there is
something in its form and mode of growth, which
at a little distance, reminded me of the dwarf Furze.
It is in general a low compact bush, but I saw a few
which, favoured by some local circumstances,
assumed the form of little trees, 5 or 6 feet high.
Mr. Smith says it is very beautiful when in flower;
at this time I could not find a vestige of a flower nor
even a pod.

The last part of the ascent to the *Cumbre* (the ridge of the mountains surrounding the Peak itself) is very steep and difficult. The bushes of Adenocarpus are thinly scattered, and almost lost amidst the wilderness of loose stones and bare rocks which give an extraordinarily savage character to this part of the mountains.

The lava (black or red, very porous and often scoriaceous, with much elongated pores, but not vitreous nor approaching to obsidian) assumes indescribably strange rugged and fanciful shapes. At length the ridge was gained, open and level spaces began to appear amidst the crags of lava, and presently without any marked descent, we entered the great elevated plain of pumice stone which extends for many square miles around the Peak. Unfortunately the mist in which we were enveloped was so thick that we could see but a very little way in any direction, and could not get even a glimpse of the great mountain. The scene immediately around us was very singular. Large bushes of the *Retama blanca*, Cytisus nubigenus, scattered at considerable distances from one another, were the only vegetation. The surface of the ground composed entirely of small loose pieces of pumice, and utterly bare of the least trace of vegetation, except the aforesaid Retama, looked like a sandy or rather gravelly desert, but here and there rugged, fantastic rocks of lava stood up from the plain of pumice like barren islets from the sea.

The Retama blanca, which seems to be almost the only plant of this great table-land, is a kind of

1854. Broom, which in its general appearance, may be said to be intermediate between the common and the Spanish Broom.   Its round (not angular) smooth, leafless twigs, most resemble those of the latter, but it is very much more branched and bushy. Its stems and main branches, covered with a smooth yellow bark, spread very near the ground, and the smaller branches rise up with a peculiar curve.   The largest and vigorous bushes are very regularly hemispherical, the smaller ones more flat topped.   It is now entirely out of flower, and I could not find even a pod that was not empty and shrivelled.   When in blossom, Mr. Smith says, the bushes appear really as if covered with snow. from the profusion of their white flowers, and the fragrance is so powerful that in the early morning it may sometimes be perceived even at the Port of Orotava.

It was but a little way below the Cumbre that I saw the first plants of Retama.   On the plateau of the Canadas I saw nothing else, except a few scattered plants of the Adenocarpus.

The clouds were so hopelessly thick, and the hour so late that we were obliged to give up all hope of seeing the Peak, and to turn back without proceeding far on the Canadas.

Night came on before we were half-way down, and our descent in the dark was very tedious, difficult and even dangerous.   We did not get back to the inn till eleven p.m., having been out fourteen hours.

Mr. Smith, who has often ascended the Peak,

and who appears to me to be an excellent observer, 1854.
quite agrees with Webb and Berthelot, that the
region of *grasses* mentioned by Humboldt, has no
existence on the Peak; that except a Violet, and
perhaps one or two other small herbaceous plants,
the Retama is the last flowering plant seen on
the ascent.

I was much interested in observing the successive
zones of vegetation in this ascent; they are
peculiarly distinct and well marked. First—the
region of cultivation ; second—the zone of the Erica
arborea, Faya and Hypericum ; third—the zone of
the Adenocarpus ; and fourth—that of the Retama.
It would have been still more striking if the large
trees which formerly occupied part of the second
region, had not been destroyed. It is difficult to
fix any limit between the region of the Cactus and
that of the European cultivation. It appears to me
very surprising that Humboldt should have omitted
all mention of the Adenocarpus, a plant so
eminently social, and occupying by itself so large a
zone of the mountain. The country name of the
Adenocarpus frankenioides is Codeso.

March 30th.

We made another attempt at the Cumbre, but
failed through the slipperiness of the road and the
timidity of our guides. We passed through Realejo,
and ascended the great mountain-wall to the W. of
it by the excessively steep zigzag road called the
Vueltas de Tigayga, by which we had returned from

1854. Icod on the 20th. Then, turning to the left just
before reaching Icod Alto, we ascended by a steep
clayey path, at first skirting the same deep barranco
which runs down between Icod and La Guancha.
This ravine at least in this upper part, is cut
through beds of soft clayey tufa, and has a different
character from any others I have seen in this island,
more resembling the waterworn gullies of Brazil.
Our path was a very fair one so long as it was dry,
but when we got up into the region of the Heath,
we found that there had been rain the night before,
and the ground was so excessively slippery as to
be very unsafe for the horses. We had, however, a
splendid view of the Peak for a few minutes before
the clouds drifted over it. The aspect of this
solitary cone, glittering with snow, so lonely and
unbroken, has something quite different from any
other great mountain that I have ever seen so near;
it is very majestic. To our left as we ascended the
long sloping ridge of the Tigayga mountain, we
had a fine view of the valley of Orotava; Realejo
lying, as it seemed, beneath our feet, and looking
like a model in miniature of a town. To the right
was seen the coast to beyond Garachico, and almost
to the western extremity of the island. We
stopped some time to botanize at Fuente de Pedro,
a pretty spring in a small ravine, situated in the
region of the Heaths, and I should suppose somewhere near the level of Agua Mansa. Here there
remain a few Vinaticos, Laurus indica, the only
trees that we saw in the region of the Heath. I
found here Aspidium aculeatum (which I had not

seen before in Teneriffe) the same beautiful Myosotis 1854.
as at Agua Mansa, and a few Mosses.

<hr />

We took a farewell of the beautiful valley of
Orotava, and returned to Santa Cruz, going by the
upper road, for the sake of seeing the wood of Agua
Garcia. The day was fine and hot and nearly calm,
but very hazy, so that we could not see very far,
and had no view of the Peak after leaving Orotava.
The abundance of Date Palms about St. Ursula is
very striking. The rich country from Orotava to
Matanza appeared still more beautiful than the
first time we travelled this road, from the Fig trees
being in full leaf. The Peach trees in blossom have
a charming effect.

From Matanza we ascended a long way into a
high, bare, rather bleak, and uninteresting country,
where (as usual on the high grounds here) corn and
lupines are the prevailing crops. Went on till we
saw Tacoronte far below us; and after travelling 4½
hours from Orotava, turned out of the road to
examine the wood of Agua Garcia. This is but a
little way off the road, to the right, but so curiously
concealed by a bend of the hills that one does not
see it till one is very near. It is a wood entirely of
evergreen trees, and certainly, though without anything of the grandeur of tropical forests, is one of
the most beautiful woods I have seen; not so remarkable for the size of the trees as for their forms and
grouping, the richness of the foliage, the freshness
pervading the whole, the picturesque disposition of

1854. the rocks and water, and the beauty of the undergrowth, especially of the Ferns and Mosses.   Ferns,
indeed, are in greater profusion and luxuriance than
I have ever seen them before out of the tropics;
they even surpass those of Madeira.   The Woodwardia and the Aspidium aculeatum grow to extraordinary size and beauty, and the Trichomanes
speciosum perfectly mantles the wet overhanging
banks with its dark-green glistening fronds.   I never
saw it in such abundance before.

Of herbaceous flowering plants, the most interesting I saw was the beautiful Geranium anemonifolium,
which grew abundantly in several moist spots.

In the short time we were able to spend in this
very interesting wood, I was not able to examine the
trees with half the attention they deserved.   The
most abundant trees appeared to be the Laurus
Indica (Vinatico), Laurus Canariensis, a holly with
remarkably large and broad leaves (I think Webb
and Berthelot call it Ilex macrophylla), the Erica
arborea, and Myrica Faya; the Viburnum rugosum,
too, may here be well called a tree.   The Erica
grows to an astonishing height; I really believe as
much as 40 ft.   Three trunks that I measured were
3ft. 4in., 3ft. 6in., and 3ft. 11in. round; but they
are always slender in proportion to their height.   Of
the trees that I saw, the largest by far were the
Vinaticos; the largest of these appeared to have
been cut down, and to have shot up again from the
stool, but some of the stumps were of great bulk.
The other trees, in general tall and slender.

I had little time to search for Mosses.   By far the

most abundant, clothing the trunks of the trees, was 1854.
Hypnum cupressiforme. There was also great
abundance of the beautiful Fissidens which I found
at Madeira (F. serrulatus, I take it), growing on the
wet banks and rocks, in company with the Trichomanes. But I saw nothing of the Bartramia-like
Moss so remarkable at Agua Mansa. Jungermanniae
are in prodigious quantity. I was much surprised to
see several Date Palms, growing, apparently in
health and vigour, though not very tall, by a house
at a very short distance from the wood of Agua
Garcia. I had by no means expected to see them
at such an elevation.

The singing of birds in this wood was perfectly
delicious. The most celebrated singing bird of these
islands is the *capirote*. I have not been able to see
it, but it is said to be a soft-billed warbler, and the
description sounds like the black-cap. In descending from Agua Garcia to Laguna, we had a good
view of a very large and remarkable crater, among
the hills N. W. of the city.

We were about eight hours on the road, exclusive
of stoppages.

———

April 4th.

We made an excursion, with C. Lyell, along the
coast to the N.E. of Santa Cruz, as far as the next
valley beyond that of Bufadera. Lyell ascertained
that the land shells which we observed on the
25th of February, in the headland beyond the
Bufadera valley, are not contained in a regular

1854 formation, but in a "talus" formed by a slip. Their date is, therefore, geologically speaking, very recent, although the mass of loose volcanic materials in which they are imbedded has had time to acquire a considerable degree of consistency. This is an instructive lesson in the art of geological observing.

In the lowest part of the sea cliff, immediately beyond the Bufadera valley, we observed a nearly vertical dyke, about a foot wide, of black lava, presenting an unusual appearance; for it is nearly compact at the sides, and very porous, even scoriaceous in the middle, whereas, in general, volcanic dykes are of a more compact texture in the middle than at the sides. Lyell supposes that in this case the lava of the sides was first formed, and the fissure widened by subsequent shocks while the process of injection was still going on.

We were struck with the difference of character of the comparatively wide and flat-bottomed valleys on this part of the coast (Bufadera and its neighbours) from the very narrow *barrancos* which are so general on the north-west coast. Lyell conjectures that the wider kind of valleys were formed partly by the action of the sea, the others entirely by running water.

Found Fagonia Cretica, a plant new to me, in flower and fruit, on the edge of the high sea cliff.

Euphorbia Canariensis now in blossom on these cliffs.

———

April 5th.

We went by M. Berthelot's recommendation, to

see a Baobab tree, Adansonia digitata, in the garden 1854.
of Sr. Mendizabal, in the town of Santa Cruz. It
is about 35 years old, the trunk, according to our
measurements, twelve feet in circumference, close to
the ground, and nine feet one inch at the height of
about four feet. As has often been remarked in this
species of tree, the height is inconsiderable in
proportion to the thickness of the trunk. The bark
is whitish and smooth. At this season the tree is
leafless, and the large gourd-like fruits, hanging in
great numbers from the bare branches, have a
curious appearance. The form of the fruit is a
lengthened oval, the colour greenish-brown, the
surface downy; it hangs by a long and comparatively slender stalk.

I have strangely omitted, in my Journal of the
20th March, to mention the obsidian, which we find
in large quantities in the wild rugged country through
which we ascended from Icod de los Vinos to the
higher region. Unfortunately, being misled as to
the length of the day's journey, and thinking we
were pressed for time, I did not allow myself time
to examine it *in situ*, but merely looked at the pieces
which were scattered over the surface, or employed
in the construction of fences. It is less transparent,
less thoroughly glassy, than the obsidian of Lipari,
and partakes somewhat of the character of pitch
stone.

———

Rubia fructicosa is a common plant in the neighbourhood of Orotava, in the hedges and thickets,
constantly mingling with the brambles. It is un-

1854. pleasant to handle, the leaves being very prickly, but is an ornamental climber, with its long red shoots, and glossy bright green leaves and pearly berries; the flowers are small, and not showy. The bramble which is so common about Orotava appears to be the same with our pink-flowered Rubus fruticosus. Rumex Lunaria abundant in many places on the coast about Orotava and Icod, especially on the rugged tracts of lava, or "malpais." A handsome shrub, with its large, shining, bright green leaves, and wide bushy panicles.

Date Palm. The scars left by the fall of the leaves are of a somewhat rhomboidal form, and though much elongated in the transverse direction, do not reach half-way round the stem; hence very unlike the completely annular scars of Areca, Cocos, and their allies. The scars (or rather the permanent bases of the Petioles) are close together, prominent and very lasting, rendering the stem permanently and conspicuously rugged. The leaf-stalks appear to decay gradually, and the dilated sheathing portion lasts a long time, making the stem appear ungracefully thickened and swollen at the top. The panicles of male flowers are of a delicate pale straw colour; the female panicles, when the fruits approach to ripeness, of a rich golden hue, the stalks as well as the fruits.

The Papaw ripens its fruit regularly at Puerto de Orotava. I did not hear of the culture of Cacao or of Indigo having been attempted in the island. The Oranges of Realejo are noted, but to us those of the Val Bufadera appeared to have the finest flavour.

I nowhere saw in Teneriffe the Commelyna, so 1854. common in Madeira, nor the Ageratum conyzoides which is one of the most general weeds in the lower part of that island. Bidens leucantha is common to both islands, but most abundant in Madeira.

[We returned home about the middle of April, and spent some weeks in London].

———

London.
May 6th.

Exhibition of French pictures in Pall Mall; an " Entombment " by Ary Scheffer, dignified and pathetic : his " Francesca di Rimini" also a striking picture.

Dinner at Mary's. Afterwards to the Phillimores; met Catty and Norah, John Alcock, Miss Shirreff and the Herman Merivales.

———

London,
May 7th.

A visit from Donne : very agreeable, as he always is. Visit Babbage : a very long talk with him; his very peculiar and remarkable cast of mind. A very pleasant family dinner party at Mr. Horner's.

═══

LETTERS.

Mildenhall,
June 11th, 1854.

My Dear Lyell,

Many thanks for sending me Hooker's letter. There may be some ambiguity, perhaps, in his way of expressing himself, but it does not

1854. appear to me that he means to go so far as you
have represented him. I do not think he contends
that Conifers are " as highly developed Dicotyledons
as any others," but simply that they are actual
and genuine Dicotyledons, and not (as some have
thought) an intermediate grade between flowering
and cryptogamous plants. I am quite willing to
admit, looking both at the structure of the flowers
and fruit, and at the mode of growth of the stem,
that they are true Dicotyledons, and it is very likely
that too much importance may have been attached
to the *gymnospermous* character, by Lindley and
others, who have made it a ground for the establishment of a separate class in the vegetable
kingdom.   The genera Ephedra (which is gymnospermous), and Casurina (which is angiosper-
mous), seem to establish a distinct connexion
between Conifers and ordinary Dicotyledons.—But
on the other hand, I cannot help thinking with
Lindley (who is surely not a " half instructed
botanist"), that it is through Conifers and Cycads
that Exogens are most nearly connected with
acrogenous Cryptogams. What can be more striking
than the similarity (external at least) of Conifers to
Lycopodiums ! Among fossil plants too it is still a
doubt whether Sigillaria and Calamites ought to be
referred to the class of acrogenous Cryptogams, or
to that of gymnospermous Dicotyledons.—Hooker
perhaps will say that these are merely external resemblances, which do not affect the question of real
affinities : and I quite admit that Conifers are not
as closely related to Acrogens as to Exogens ; but

still these striking analogies appear to me very well-worthy of attention. I know of no such analogies between Acrogens and any part of the Monocotyledonous class.

It may be a difficult matter to say what is to measure the rank, or degree of development of different families of plants : but I do not at all see how plants that have neither calyx, nor corolla, nor seed-vessel, nor spiral vessels, nor ducts, (as is the case with the Conifers) can be considered as a highly developed form of flowering plants.

I may sum up my notions on the subject thus :— I believe that Conifers are true Dicotyledons, but Dicotyledons of rather a low grade of organization, and presenting some striking analogies to the Acrogenous class ; and I see no reason for ranking them above Monocotyledons *in general*.

With respect to Grasses, it is quite possible that their leaves may occur in the Coal formation ; but I have never seen any.

One thing more I would add by way of caution. Do not in geological theories trust too much to vegetable fossils ; I fear you may sometimes find them almost as unsafe supports as an Austrian alliance.

We spent two very pleasant days at Barton. Surprisingly little damage has been done there by the severe winter and spring ; *here* a good deal more.

With much love to Mary, believe me ever,
Your affectionate Friend,
C. J. F. BUNBURY.

Mildenhall,
June 30th.

My Dear Lyell,

I am much obliged to you for sending me Hooker's letter (or rather essay), which is very instructive and very valuable. I reckon him a *first-rate* authority in botany, and everything that he says worthy of most serious attention; but I do not consider even him infallible, and I have a few observations to make on what he says, as well as some corrections of my own former remarks, which I will put down in order.

1. I believe that I was mistaken (misled chiefly by Lindley) in attaching importance to the purely *external* resemblances between Lycopods and Conifers. At the same time I think it is in general much more difficult to distinguish between relations of *affinity* and of *analogy*, in the vegetable than in the animal kingdom; for the very same reason which makes it difficult to decide upon the relative *dignity* of different families of plants; namely, that the relation of structure to functions is much less distinct and decided (as Hooker observes) in plants than in animals. Still I ought to have remembered that the resemblances between Conifers and Acrogens are merely external, and do not in the least invalidate the broad and absolute and most important difference which exists between them in the structure of their fructification. I now fully admit that Conifers are, to the full extent of the term, *flowering* plants, and not to be considered as in any way intermediate between the two great classes.—

2. I think that Hooker underrates the degree of 1854. external resemblance between Lycopods and Conifers; though I admit that this resemblance is of no systematic value. The doubts and differences of opinion that have taken place among botanists as to *which* of these two families various fossil plants should be referred to, show, I think, that they may assume very similar outward forms.

3. I have no copy of my letter by me, but I suppose I must have said that I did not know of any such striking analogies to Cryptogams among Monocotyledons, as I supposed gymnospermous to present. If so I certainly overlooked Lemna which has very much the look of a Cryptogam. But as Lemna is merely a low form of a well-developed family (Aroidæ), and as its resemblance is to one of the *lowest* types of Acrogens (namely Riccia) no one could be deceived into supposing that *this* resemblance indicated any real passage.

4. I do not at all know on what ground Hooker maintains that " the duct or spiral vessel is a far " lower organization than the woody fibre with " discs." As far as I know, botanists are perfectly in the dark with respect to the functions or uses of those curious glandular discs which occur on the woody fibre of Coniferæ ; and therefore I do not see how we can argue that they are a high or a low structure. As far as elaboration, or complexity is concerned, I do not see that the woody fibre with discs is more elaborate than the scalariform vessel. That this peculiarity is not of great systematic importance, I should be inclined to infer from the

1854. fact that the genus Drimys, which is undoubtedly Magnoliaceous, has its woody fibre furnished with discs like those of Coniferæ, while Magnolia itself, and the Tulip tree, are without them.

5. I did not clearly express my meaning with respect to the (supposed) absence of spiral vessels from Conifers. I did not mean that the want of spiral vessels *in itself* argued a low development, but that Conifers appeared to have fewer varieties of tissue, and in so far a more simple structure than ordinary Dicotyledonous trees. The stems of ordinary Dicotyledons are composed of at least four varieties of tissue — cellular tissue, woody fibre, dotted ducts and spiral vessels. The wood of most Conifers is composed of the two former only. I forgot however at the moment, that Abies Douglasii has *apparent* spiral vessels; but I rather think they do not unroll. Probably we ought not to build much upon a character which thus appears to be present in one species and wanting in others of the same natural genus.

6. (In answer to your question) : I believe that Taxus has not real spiral vessels ("*trachées deroulables*"), nor yet strictly scalariform ones ; but the vessels of its wood are marked with spiral lines *as well as* with discs.

7. I am not satisfied by anything Hooker says that Exogens are absolutely higher in the scale of creation than Endogens. It appears to me that the characters by which flowering plants are separated from flowerless are altogether of a different nature, of a different order of importance

from the distinctions between Exogens and Endo- 1854.
gens. All Phænogams have stamens and pistils,
organized essentially on one type (as Robert Brown
first showed) and are propagated by seeds, which are
developed through the mutual action of the stamens
and pistils, have a distinct embryo, and germinate
in a definite manner. Cryptogams have, properly
speaking, neither flowers nor seed; where they
have organs which appear to represent stamens and
pistils (as in the Mosses), these do not appear to be
formed at all on the same type, their spores differ
essentially in structure from seeds, and indeed
appear to approach much nearer to the nature of
pollen grains. The most simple Exogen or Endogen
may fairly be said to be of a higher organization
than the most developed Acrogen, because the one
has perfect flowers and seeds and the other has not.
But I can see no difference, at all corresponding to
this, between Exogens and Endogens. These two
classes, it appears to me, have all the same organs,
only differently arranged. The stamens, pistils and
seeds of Endogens appear to me as perfect as
those of Exogens. In short, I cannot yet see any
reason why they should not be considered as two
nearly parallel (though not co-extensive) series,
both beginning at the same distance from Cryptogams; quite distinct, but " *magis pares quam*
similes." Of course, in books and practical
arrangements, one *must* be put before the other,
just as according to the proverb, " when two men
ride on one horse one must ride behind."

8. I quite agree as to the difficulty, or perhaps

3 Y

1854. impossibility, of deciding to general satisfaction what is high or what low among flowering plants. Almost every botanist has different notions on these points. I have been hitherto inclined to think that as a general rule, plants should be ranked higher in the scale in proportion as they have a greater variety of distinct organs in a well developed state; but I must admit that, as Hooker says, plants of a very simple and apparently imperfect structure are often closely connected by indisputable affinities with others of a very high degree of development. Hooker certainly makes out a very good case for his Conifers.

9. I was astonished at Hooker's saying that an oyster is a higher type than an insect, and I am glad to find you dissent from him.

10. That Cycadeæ have been proved to be decidedly exogenous, was quite new to me, and I am very glad of the information.

My letter is running to an unconscionable length, and I will conclude this part of it by saying that I am quite ready to admit that Conifers, though not among the highest forms of Dicotyledons, are by no means among the lowest; and this will probably be sufficient for your purpose. Of Cycads I have nothing to say, as I have not had the opportunity of dissecting any; my knowledge of them beyond mere external characters, is derived wholly from books. Pray look at what Forbes says on fossil plants, in his anniversary address, pages 70 and 71; it is admirably sound and true; and I do not say this on account of the compliment to myself; but I

most thoroughly concur, in what he says of the 1854.
rashness and unsoundness of much that has been
done by palaeobotanists.

Poor Lady Bunbury has been, and I fear still
is, very seriously ill: not in actual danger, but in
a state to make one very uneasy.

Both Mrs. Horner and Mrs. Power are very kind
and very pleasant.

With much love to Mary,

   Believe me ever,

     Your affectionate friend,

       C. J. F. BUNBURY.

―――――

         July 13th, 1854.

My Dear Father,

  I have now read through the *new* part of your
book, that is to say the introduction, the year 1800,
and the Egyptian campign, and I must write to
express my hearty admiration of it, and the very
great pleasure and interest with which I have
read it. I have always considered your Maida
campaign, the very perfection of a military narrative,
but I think the Egyptian campaign falls short of it
only in so far as it wants something of that peculiar
freshness—that zest as it were—which of necessity
belongs more particularly to the story of what you
have yourself seen. The narrative is beautiful
from its singular clearness and unaffected simple
vigour, and the style is to me particularly agreeable.
I am sure it must command the admiration of all
who are qualified to judge. I have been reading it

1854. aloud to Mrs. Power, who admires it as I do, and is particularly struck with your impartiality.

Believe me ever,

Your affectionate Son,

C. J. F. BUNBURY.

Mildenhall,

August 3rd, 1854.

My Dear Lyell,

Hooker told me of the recent discovery of numerous insects in the coal of Saarbrück, and that some of their wings had been mistaken for Ferns ! but he seemed to suppose that these were the *first* insects ever discovered in that formation, which is certainly a mistake. I see you notice in your Manual, though slightly, the insect remains found at Coalbrook Dale, and in Bohemia. Since such delicate things as insects have been preserved, we must certainly expect to meet with recognizable remains of Dicotyledonous plants, if any existed. I had some good conversation and discussion with Hooker, though several things now occur to me which I wish to have shown or mentioned to him, and which I forgot. He was much struck with my specimens of Equisetites from the coal formation of Cape Breton. He explained to me better than I had understood it before, the structure observed by Mr Williamson, in the Calamite; and his own reasons for still thinking it a very highly developed Cryptogam, reducible to the Lycopodium type of structure, rather than an Exogen. I own I am not

quite satisfied, though I see now that the medullary 1854.
rays are not as certainly characteristic of Exogens
as I had imagined. I think, at any rate, that
Calamites (including Calamodendron) and Sigillaria
were very remote from any plants now existing. In
my book I mean to put them, with various other
things, in a group by themselves, under the head of
Doubtful Cryptogams, between the decided Cryptogams (which will end with Lepidodendron) and the
decided Exogens ;—that is to say if I am still of the
same way of thinking by the time I get to them.

My palaeobotany, in truth, is a work like
Penelope's web; every time I go back to my
M.S.S., after some length of absence, I find so
much to correct, to alter, to add and to retrench,
that the end seems as far off as ever. There are
some points on which my opinion has changed
three times since I began. I get on more flowingly
with my Madeira notes, and Hooker was glad to
hear I was engaged on that subject. How much
more agreeable it is, writing of what one has one's
self seen, than compiling! By the way I have
undertaken at Arthur Hervey's particular request
to enlighten the good people of Bury upon the
wonders of Madeira and Teneriffe. Peacock said
he had heard great things from Mr. Vernon Har-
court, of your Madeira discoveries.

Do you know that Hooker, like myself, is sceptical
as to the enormous longevity attributed to Adansonia ? Our doubts are grounded on the rapid
growth of the tree, and its loose, spongy texture.
We had four most delightful days with Hooker,

1854. Sedgwick and the Smiths. I am positively in love with Mr. Smith ;* he is one of the most agreeable men I have ever met. We had a capital day at Ely. Fanny was obliged to take a good rest next day, but she is very well, and has been hard at work to-day at the schools, with the inspector.

Ever yours affectionately,
C.B.

―――――

Mildenhall,
August 17th. 1854.

My Dear Mr. Horner,

I owe you many thanks for your interesting letter of the 6th, from Berlin, but I fear I have little of interest to send you in return since Fanny has given you a full account of our delightful visit from Sedgwick, Hooker, and the Smiths. All the accounts we have received from Berlin during your visit there, have been extremely agreeable, and give us a very lively and pleasant idea of dear Leonora's† household and belongings. I am very glad you have been spending your time so much to your satisfaction, and have met so many remarkable and interesting men.

I look forward with great pleasure to our intended visit next summer to so great a centre of scientific, literary and artistic activity and progress as Berlin, and hope to become acquainted with its celebrities. But above all I am desirous to see and hear Humboldt, whose writings have so long been my delight. I am now reading his " Ansichten " (in

* Samuel Smith, son of Wm. Smith, M.P. for Norwich and Uncle of Florence Nightingale.
† Married in May of this year to Chevalier Pertz, principal Librarian at Royal Library at Berlin.

German) with great enjoyment, and do not find it 1854.
at all difficult. I really hope in time to gain a
tolerable *reading* knowledge of German; more I do
not expect. Your account of Ehrenberg and his
discoveries is very curious and satisfactory; I had,
I confess, been inclined to suspect that his microscopic organisms were somewhat like Bowerbank's
sponges, and so Hooker thought; but you seem to
have satisfied yourself that his observations may
really be depended on. It would seem as if those
minute creatures, animal and vegetable, peopled
those parts of the world which from depth of water
and other causes are uninhabitable by anything
else. Is not their existence in the oldest
sedimentary rocks (otherwise unfossiliferous), rather
favourable to the views of the *progressionists?* Or
must we suppose merely that those deposits were
formed in very deep seas?

I have seldom passed four more agreeable days
than from 27th to the 31st of July. Joseph Hooker
really deserves to be called the English Humboldt,
for the extent and variety of his scientific knowledge.
Sedgwick was in high force, excellent both in his
humorous and serious vein; and I know few men
who suit me so entirely or who are in every way so
agreeable as Mr. Sam. Smith. We have since had
a pleasant visit from my Father and Lady
Bunbury, and we expect Mrs. Jameson to-morrow,
and are going to take her over to Barton.

My reading has been rather various this summer:
Hooker's "Himalayan Travels," my Father's new
book, and "Southey's Life."

1854. There is a masterly article in the late number of the
Quarterly, on "Lord Lyndhurst (or rather indeed on
Lord Aberdeen) and the War :" a merciless
dissection of Lord Aberdeen's weak and shilly-shally
policy or impolicy: and the more effective, because
perfectly courteous, temperate, and gentlemanlike in
style. I should like very much to know who is the
writer. He is of opinion that there are no symptons of
dissolution in the Ottoman Empire, nor anything to
forbid the hope of a healthy though gradual progress.
With respect to German politics I will say nothing
further than that I wish our Government would not
allow themselves to be hampered and held back
by false friends or slippery allies.

"Da chi mi fido guardami Dio ;
"Da chi non mi fido, mi guarderò io !"

We are still quite in the dark as whether the
great combined expedition of which the *Times*
has been talking so much, has really sailed from
Varna or not, and if so what is its destination. As
Henry is certainly in it (supposing there *be* such an
expedition), we cannot but feel very anxious ; but at
the same time, I feel very sure not only that Henry
will do his duty perfectly well, but that there is
nothing he so much desires as to have the opportunity of doing his duty on active service, and few
things he would less desire than the inglorious
forced idleness which has hitherto been the lot
of our army in Turkey. I am glad our Ministers
had the grace to pay a deserved compliment in the
Queen's speech, to the valour of the Turks. Whatever the Black Sea campaign of our army and navy

may turn out, the parliamentary campaign has not 1854 been brilliant. I hope Lord Raglan may carry more Russian positions than the Ministry have carried measures.

I wish we had any chance of seeing you here this autumn. Do come to us if you can manage it. Much love to Mrs. Horner, Susan, and Joanna.

Ever your affectionate Son-in-law,

C. J. F. BUNBURY.

Mildenhall,
August 31st, 1854.

My Dear Katharine,

I am very much obliged to you for your kind offer of the Indian woods, which I shall be very glad of; they will be welcome additions to my collection.

We have had some delightfully warm days, and have enjoyed sitting out on the grass, under our great trees, and in full view of our flower beds, which are in great beauty. The trees, which Mr. and Mrs. Horner are so vehement against, and which certainly do give us too much shade in damp and cold summers, are delightful in such weather. The harvest has been extremely fine here, and, for a wonder, the farmers appear for a time to be really well satisfied.

I have been very much interested in "Lorenzo Benoni."* It is a very well told story throughout, and a curious and striking picture both of Italian

* Written by an Italian Exile—Ruffini.

1854. governments, and of the proceedings to which generous and noble-minded men are driven in such
countries, by the want of the power of free and open
opposition. The story of his escape is singularly
interesting. I should like much to know who the
author is; he must be a man of extraordinary ability,
and his English is beautiful. I have also just
finished the life of Kossuth, which is well done, and
has given me much new information; in fact, I never
before well understood the rights of the Hungarian
question. Kossuth is certainly a man of wonderful
eloquence and energy, and of great sagacity; and his
speech the other day at the Potteries was a remarkably powerful one, though there was much in it that
I did not like; but I do not think our Government
can fairly be blamed for not following his guidance.
His paramount object naturally is to stir up enemies
from all sides against Austria, with a view to the
liberation of Hungary.

Austria is to him what Rome was to Hannibal;
and no one can justly blame him for this. But it
does not follow that other nations ought to make
their policy subservient to his views. At the same
time, I think our Government very much in the
wrong in allowing their measures to be so much
influenced and hampered by Austria, from which, I
am satisfied, they will never receive any support in
this war that will be at all an equivalent for the
mischief of delay.

I see the Czar tells his subjects at Odessa that the
Austrians are entering Wallachia to keep the Turks
out of it! This is exactly what Kossuth said, and I

was much inclined to believe it true; but since the 1854.
Czar tells his subjects so, I conclude it must be false.
Still, I do not at all believe that they (the Austrians)
mean to do anything more than quietly to occupy
Wallachia—which may be very convenient for them.
As for the Prussian Government, I have no doubt its
object is (and has been all along) to give Russia all
the support and assistance it can, without involving
itself-in war. The Bomarsund affair, though a small
victory, was neatly managed, and is encouraging, as
it seems to show that the granite batteries are not as
impregnable as they appear. I hope it will soon be
followed up by a greater stroke.

I am very anxious about my brother Henry;
though not actually ill, by the last accounts, he was
very weak, and out of spirits, and in that state of
physical and moral depression the maladies of the
climate and of the season are much to be dreaded.
I fear. much more for him from sickness than from
the Russians. The troops generally, indeed, seem
to be in the like state of depression, and the continual
delay of the long-promised expedition must be very
trying to their spirits. I do hope something will be
done in earnest very soon.

I shall be exceedingly glad to see both of you
here, and hope you will stay a long time.

Ever affectionately yours,
C. J. F. Bunbury.

1854.   *From his Father.*

Barton,
September 24th, 1854.

My Dear Charles,

It has indeed been a great relief to know that our army has been landed without any of the loss which there was reason to dread if the debarkation had been made in the face of the enemy. On the other hand there is this disadvantage in going to Eupatoria, that our troops will have to march five or six days through an open country which is very favourable to the Russians, who have probably a strong force of cavalry, while we can have little more than a thousand light horse. There is likely to be hard fighting before we get within cannon shot of Sebastopol. Our latest account of Henry is in a letter of the 6th, from Harry and his mother. He had seen his Uncle that morning, and reports him to be better, and in high spirits at the prospect of a battle. We know, however, that as late as the 3rd, Henry had not got quite rid of the diarrhœa.

The affectionate attention of young Harry to his Uncle is very gratifying. Lord Edward Russell also speaks highly of him.*

We still hold the intention of going to Felixstowe next Wednesday, if we hear from Henry after his landing, by that day. If not we shall probably wait till his letter arrives ; and you shall have further notice, for I trust you will not relinquish your plan of joining us.

Ever most affectionately yours,
H. E. B.

* Young Henry Bunbury, the Son of Hanmer Bunbury.

From his Father

Felixstowe, Ipswich.
October 1st, 1854.

My Dear Charles,

Cissy has received letters this morning from Henry up to the 17th September. Poor fellow he had relapsed after embarking, and the diarrhœa soon passed into dysentary, and he had evidently been in great danger. The disease had happily given way, and he hoped, had left him altogether. But he was reduced to such a state of weakness that his landing with his regiment was impossible ; and he had, with what deep reluctance you will well comprehend, consented to be sent back, with very many other sick, to the Bosphorus. He trusted that the air and quiet of Therapia would restore his strength, and that he might be able to rejoin the army before the siege of Sebastopol could be over. But it is impossible not to see that his health has been fearfully shaken, and we can only wait with more anxiety and submission for further accounts. What Henry says of the debarkation of the army agrees with what we read in the *Chronicle*. Will. Codrington,* who had succeeded to the command of the Fusilier Brigade, had come immediately to Henry, and had shown him great cordiality and kindness. Poor Cissy bears up wonderfully well, and she tries to see everything in the most favourable light.

This is a nice little place, and I wish very much

* General Sir William Codrington.

1854. you would come to us; it would do you good, and your coming would do us good.

Much love from us all to Fanny,

Ever most affectionately yours,

H. E. B.

———————

My Dear Charles,

I had meant to have written you a few lines to thank you for your affectionate letter; but Cissy has put everything else out of my head by communicating to me the news she has received this morning from Mildenhall.* Most highly gratifying to my heart is this intelligence. I know it will add greatly to your own, and to dear Fanny's happiness, and warmly do I congratulate you both.

Do not think of coming to Felixstowe; Emily and I shall probably be at Barton on Tuesday, and Cissy on the following day. I will write to you again. I sent you my speculations as to the Crimea yesterday. I had been from the first an unbeliever as to the blowing up of Fort Constantine, and the surrender of Sebastopol. But we are on the right side of the place now, and I am now very sanguine as to the result. I wish I could be equally sanguine as to our dear Henry's health. God preserve you and Fanny. I pray that you may have a son who may be as great a comfort to his father as you are to me.

Ever most affectionately yours,

H. E. B.

* This letter refers to the expectation of a family, which, however, was disappointed.

Mildenhall,   1854.
October 8th.

My Dear Mrs. Horner,

I thank you very much for your kind letter,
as well as for your pretty present of the bronze leaf
inkstand, for which I should have thanked you much
sooner if I had known that it was yours. It is very
pretty and very useful.—Our successes in the Crimea
have been glorious, but purchased at such a sad cost
of valuable lives that one can but half rejoice in them.
The terrible loss of officers, especially in the 23rd,
shows how hard a battle that on the Alma must have
been. It makes me sick at heart to think how many
fine young men have fallen in their first battle, and
how many families are mourning for those dearest
to them.

Our poor friend, Mrs. Evans, has lost her brother,
Lady Walsham has lost a son, the Waddingtons
have lost a nephew. Yet with all this, I cannot
thoroughly rejoice that Henry was not in the battle,
for I feel how deep and cruel a mortification it will
be to him to have missed his share in such gallant
actions, and I fear disease for him even more than
the chances of battle. There has been no further
account of him since the 17th. It is a most trying
time, especially for poor Cecilia, and it is a great
blessing that she has so much firmness and self-
control.

Dear Katharine's children are very pretty, amusing little creatures, especially Frank, and they and
my little niece and nephew, Fanny and Herbert,
seem very happy playing together in the garden,
which is indeed a pleasant and secure place for them.

1854. Susan is painting some beautiful illustrations for my intended lecture at Bury.

I believe I shall go to-morrow to Felixstowe for a few days. I am very unwilling to leave Fanny, but she feels that it is my duty to go to my father to whom I may be some comfort. She sends her affectionate love.

With my love to Mr. Horner and Joanna, believe me,

Your affectionate Son-in-law,
C. J. F. BUNBURY.

To Lady Bunbury.

Mildenhall.
October 10th, 1854.

My Dear Emily,

I cannot express to you how much I was touched and gratified by the warm affection and sympathy of which your and my dear father's letters of Saturday are so full. I feel your kindness from my very heart, and so does Fanny; and indeed my dear *second mother*, I have always been very sensible of all the affection and goodness you have so constantly shown me.

Fanny seems very well on the whole, but it is difficult to keep quiet in these times of such extraordinary excitement. What a time it is indeed! and with what a mixture of feelings one reads the despatch of the battle : joy and pride in the glorious deeds of our countrymen, and sorrow for so many gallant men who have fallen, and for so many families that are mourning for their bravest. What

a tremendous loss of officers, especially in the 23rd. 1854.
I suppose there was hardly ever a more brilliant
exploit than that on the Alma.—It was an in-
expressible relief to me this morning, to learn from
my father's note that dear Henry had reached
Scutari alive, and was getting better, though his
recovery seems to be slow. I had been very nervous
about him. I trust he will soon gain strength, but
that his impatience to be in action will not hurry
him back to the Crimea before he *has* gained enough
to bear such rough work. Much as I have felt for
his disappointment, the feeling of thankfulness that
he was *not* there almost predominates again when I
look at the fearful list of killed in his regiment, and
think that, in all probability, he would have been
added to the list. I feel very much for our kind
neighbour, Mrs. Evans; a warm-hearted creature
she is, and the common interest of having brothers
in the same regiment, has been a great link between
us.

The poor Walshams, too, are deeply to be
pitied. Poor Arthur seems, by all accounts, to have
been a very fine fellow. What a number of fine
young men have perished in their first battle! But
when one considers that most likely nine-tenths or
more of our troops had never seen an enemy before,
it was really a wonderful fight. I was very much
interested by my father's military remarks, and long
to hear his further criticisms, especially on the battle
of the Alma. It appears from the latest accounts
that the Russians made no stand on the Katcha;
indeed, I daresay they were a good deal dismayed at

1854. being so quickly beaten out of their formidable position on the first river.

I long to see you all, and I propose to come to Barton to-morrow, and stay with you a day or two, as Fanny will not mind parting with me for so short a distance. I shall leave her in excellent hands, and indeed the presence of her dear sisters is an immense comfort to me.

Being obliged to go into Bury yesterday about some gaol business, I went on (with Katharine Lyell) to call at Hardwicke, and found Lady Cullum full of warm feeling and anxiety about Henry, as well as sorrow for the Walshams. I do not write to my father to-day, for I hope to see him to-morrow, and I decided to write to you this time, as I have written more often to him; but I assure you that nothing ever went more to my heart than the last lines of his most truly affectionate letter. It is my fervent prayer that God may long preserve both him and you, and grant you both every blessing of an honoured and happy age.

Ever yours most affectionately,

C. J. F. BUNBURY.

Mildenhall.
November 6th, 1854.

My Dear Mr. Horner,

I must congratulate you on having dear Susan with you again, and dear Katharine and your grandchildren in your neighbourhood. I need not say how great a pleasure their visit has been to us, or how much we miss them. You will have heard

from Susan that my lecture at Bury* was as successful as I could have desired, and almost more than I
could have expected. It certainly was very much
indebted to the really beautiful illustrations which
she had so kindly taken so much trouble to paint for
me; I cannot tell you what service they were of, or
how much they were admired. I had a very attentive
and favourable audience, and a very numerous one,
—more than 600 persons, it was reckoned,—which
perhaps is more than you would have expected in a
small country town like Bury. It seemed odd the
holding forth in the ball-room where I had so often
danced, in my younger days (and where Mrs. Horner,
too, I think has danced); but it makes a first-rate
lecture room. The only misfortune was that my
matter was much too copious for the time I could
allow myself, and from want of experience, I could
not rightly proportion it: consequently, as I went on,
I found time gaining upon me so much that I was
obliged to leave out much that I had intended to
say, and Madeira, coming first, received more than
its due share of attention. In truth, the subject
would have required, to do it justice, two or three
lectures instead of one. It was an easy subject for
me, as I had the whole fresh and full in my mind,
but otherwise, not a very advantageous one for a
lecture, as being too wide and vague, and wanting
unity. Nevertheless, I am pleased to have so far
succeeded in my first attempt. Now, having got
that off my hands, I shall have more time for purely
scientific researches.

* On Madeira and Teneriffe.

1854.   I shall be thankful for any scientific news you can send me, and, pray bear in mind, that I live here *in the desert*, and am thirsty for knowledge. We have good accounts of my brother Henry, down to the 15th of last month; he was rapidly recovering strength; and what I think also good, the surgeons would not yet let him rejoin his regiment. There was nothing I was so much afraid of for him as his joining before his strength was sufficiently restored. What an anxious time it is! and how long it seems before one gets any authentic intelligence. The siege is a formidable enterprise, in the face of such numbers; yet, to such men as the British and French troops showed themselves at the Alma, everything seems possible, and Henry writes that Lord Raglan is said to be in high spirits, and confident of success. All the details are interesting in a high degree, especially the private letters of soldiers as well as officers, which appear in the newspapers. There is a union of domestic affection and tenderness of feeling, with heroism, which is very admirable. And then how beautiful is the self-devotion of Florence Nightingale! She is certainly a most remarkable and admirable person.

With much love to Mrs. Horner, Susan, and Joanna, and *from* Fanny to all of you, I am ever

Your affectionate Son-in-law,

C. J. F. BUNBURY.

Mildenhall,
November 19th, 1854.

My Dear Katharine,

I was very much obliged to you for your pleasant letter of the 17th, and very glad to hear such good accounts of you all. We are in painful anxiety about my brother, as in all probability he must have been in that terrible battle of the 5th, and we do not know whether his regiment suffered much or not; all the accounts, and especially the number of general officers killed or wounded, show what a close and terrible fight it must have been. In a very few days now we must have the returns, and one can only hope and pray that Henry may have come off safe. Cecilia's fortitude, prudence, and self-control, are admirable. She has a letter from him of the 2nd, at which date, though not very strong, he reports himself as *not ill;* but there was much sickness in the army, I suppose from over fatigue, as he says the cholera had ceased, and the provisions were very good.

But his view of the progress and prospects of the siege is very gloomy. It is, indeed, a formidable enterprise, and I should think one of the most daring things ever undertaken on a great scale in war. The enemy prove to be very superior to us in number and weight of guns, and efficiency of fire; and if their infantry were as brave as ours, we should have little chance. However, Henry still thinks that the town will be taken, and if it be, it will then be in our power to destroy the ships and dockyards, and the rest of the naval establishments, which are all on the

1854. south side of the harbour, and the destruction of
which is the great immediate object. At any rate,
every nerve must be strained to succeed in the
enterprise, which has already cost us so much blood;
and it will be monstrous if England and France
together are not able to cope with Russia.

Frederick and Augusta Freeman have been with
us now five days, and most pleasant and cheering
their company has been; *she* is quite charming, and
I like *him*, too, very much. They leave us, I am
sorry to say, to-morrow.

I am glad you have had some pleasant botanical
talk, and that you like the ferns I put up for you. I
have not done much since you went, besides revising
and correcting my botanical paper on Madeira, and
making out some of the Mosses and Lichens I
collected there. My time has been a good deal cut
up, especially the week before last, when I spent
much time in the plantations, making a regular
thinning. I found the Pteris aquilina growing to an
extraordinary height in one plantation. I carefully
measured one frond, which was eleven feet and an
inch high.

Give my love to Frank and to Leonard, and do
not let them forget Mildenhall.

With much love to your husband, and to the family
party at Q. R. W. and Harley Street, I am ever

Your affectionate Brother,

C. J. F. BUNBURY.

To Lady Bunbury..

Mildenhall,
November 19th, 1854.

My Dear Emily,

I thank you very much for your letter. One can hardly think of anything at present but this tremendous siege; one of the most daring and formidable enterprises, surely, that a British army was ever engaged in. Henry's last letter (of the 2nd) of which dear Cissy has just sent us an account, is tolerably comfortable as regards himself, but gives a very discouraging view of the situation and prospects of our army; so much reduced in numbers, and over-worked in consequence; and unable to match the enemy's fire. If the Russian infantry were as good as their artillery, it would be a very gloomy prospect indeed. However, it would seem that considerable reinforcements are on their way out and indeed, that five thousand French must have arrived very soon after the battle. Dear Henry, we are continually thinking of him; a very few days now must bring us the returns of loss in the battle of the 5th, and one can but wait in hope and resignation, and trust and pray that he may have been spared to us. The 23rd being reduced to such a scanty number is hardly likely to have been put prominently forward in the action. Cissy's calmness, trustfulness, prudence and self-control, in this trying time are most admirable.

The society of Frederick and Augusta is exceedingly cheering, and quite a comfort to us during this suspense, the more so as both take the brightest

1854. and most cheerful view of everything. Augusta's bright, beaming face, so handsome and so happy, is in itself quite a sunshine in the house. She is perfectly charming. I am sorry to say they leave us to-morrow.

The weather was very unfavourable during Willie Blake's visit, but nevertheless he seemed well satisfied with his sport. He is a nice quiet gentlemanlike fellow. I hope the weather allows you to take exercise, and to enjoy the beauty of Abergwynant, and that you will by degrees find your health improved by it. At this season you will hardly find anything in the botanical world there, except Mosses and Lichens, but it is a fine place for them; I wish I could be there to help you to collect and examine them; and I am sorry that, when you were here, I forgot to show you Hooker's "British Jungermannias," the standard work, (and a very beautiful one), upon a family of plants for which I think you have a predilection.

Fanny is well. Pray give my best love to my Father, and believe me ever,

Your very affectionate,

C. J. F. BUNBURY.

Mildenhall,
November 22nd, 1854.

My Dear Mr. Horner,

I thank you very much for kindly sending me the very sad news of poor Edward Forbes's death. Most truly lamented it is, both in a social and scientific view. To science the loss will not be

easily repaired, and it is particularly sad that he 1854.
should have been snatched away at the very beginning of such a career of usefulness as his appointment at Edinburgh promised. Both as a zoologist
and geologist, his acquirements were I suppose,
of the first order, and there were few if any naturalists of our time, who had a more philosophical mind,
or wider and more comprehensive views ; he seemed
to have in a remarkable degree, that rare and
valuable union of large powers of generalization with
patient accuracy and power of minute research ;
so that one cannot help feeling that, great as were
the services he had already rendered to science,
much greater yet were to be expected from him.
And then he was so good and amiable and such a
pleasant member of society, with such a variety of
talents and accomplishments, so much cheerfulness
and such an entire absence of vanity or jealousy ;—
the loss is indeed deplorable. Poor Mrs. Forbes
how deeply she is to be pitied. It is singular that
within this very year, poor Forbes should have
pronounced an eloquent eulogium upon Mr. Strickland, cut off at nearly the same period of life. As
you may well suppose my thoughts are mainly
engrossed at present by anxiety about my brother
Henry : and the suspense is most painful till we get
the returns of loss in the terrible battle of the
5th, in which he was no doubt engaged. The
slaughter of officers seems to have been fearful,
especially in the Guards, but the Russians have had
a severe lesson. But even if Henry should, as I
trust, have come safe out of the battle of the 5th,

1854. our anxiety is only adjourned, for the seige is evidently a most formidable undertaking, the success uncertain, and the cost of life certain to be much greater yet before all is done.   One can only hope that sufficient reinforcements may arrive in time. There is an excellent spirit in this country, and no symptoms of flagging or shrinking on the part of the people.   The whole question is, as the great Duke said at Waterloo, " who will pound the longest ? "

Fanny keeps well.   Much love to all your party,

Ever yours affectionately,

C. J. F. Bunbury.

———

Mildenhall,
November 21st. 1854.

My Dear Leonora,

I hear that you wish for a botanical letter from me, and I will try to write you one, though I am afraid I have nothing very interesting to tell.   I am very glad you do not intend to give up botany, though I daresay you have not as yet much time for it; but, with so much knowledge as you have already acquired, and an excellent collection of your own, and such great resources within your reach, it would be a thousand pities to drop it.   Are you acquainted with Dr. Klotezsch ?   (I am not sure whether I have put the right quantity of consonants into his name);—I believe he is a very good botanist, but the deaths among German botanists, in the last two years, have been very numerous.—I have had so many important additions to my collections this

year, that my time has been much occupied in 1854.
arranging and studying them; and I have a great
deal of pleasant work of that sort yet before me.
First, our own collection made in Madeira and
Teneriffe, including nearly forty Ferns; secondly, a
very large set of New Zealand plants, from Joseph
Hooker, including an immense number of Ferns;
thirdly, a very fine set of Ferns from Simla, which
Katharine was so kind as to give me. Altogether, I
believe that considerably more than 100 Ferns have
been added to my collection this year, and this has
set me again upon making a particular study of that
family of plants. Very beautiful and interesting
they are; but very puzzling; horribly difficult to
satisfy one's self either as to the value of the generic
characters, or as to what really are species and what
varieties. Indeed, the nomenclature of Ferns, as to
the genera, is becoming a perfect chaos, which needs
a new Linnæus to set it to rights.

Of the 40 Ferns that have been found in Madeira,
from 15 to 18 are British species; I say from 15 to
18, because in some of the cases there are differences
of opinion as to the identity.

(November 27th). Since I began this letter, we have
had the comfort of hearing of my brother Henry's
safety; an indescribable relief, as we had been in
most painful anxiety about him ever since we heard
of the terrible battle of the 5th. I have had a letter
from him of the 7th; he was in the battle, and under
a tremendous cannonade, but, by God's mercy, came
off unhurt.—On the other hand, I have been much
shocked and grieved to hear of the death of Edward

1854. Forbes.   It is an immense loss to science; he was certainly in the first rank of naturalists.   In a different way, Lord Dudley Stuart is also to be regretted.   He was a very good man, and though I think by no means of first-rate abilities, was very earnest, single-minded, and courageous in supporting what he thought the right.

(December 2nd).   At the urgent request of the Mildenhallites, I have repeated here my lecture on Madeira and Teneriffe, and again with success.

My paper on the vegetation of Buenos Ayres (which I showed you in manuscript two years ago) is now printed in the " Linnean Transactions," and I will send you one of the separate copies by the first good opportunity.   The most interesting and important thing in botany that I have read for a long time, is Joseph Hooker's " Introduction to the Botany of New Zealand."   It is a most admirable treatise.   What a remarkable man Joseph Hooker is!   I call him the English Humboldt.   This narrative of his Indian travels, contains a prodigious quantity of information, and is very pleasantly written.

By the way, Henry has sent me some dried flowers and bulbs of a beautiful autumnal Crocus, from *before Sebastopol.*   Interesting from their locality and above all (to us) as having been gathered by him under such circumstances.

There is an excellent spirit in this country, if it be but properly seconded by the Ministry; in fact the English blood is up, and I think no sacrifices will be grudged to maintain worthily our righteous cause.

I think too, there is an increasing feeling that those 1854.
who are not with us are against us, and that it is
better to have open enemies than concealed ones.

I am ever your affectionate brother,

C. J. F. BUNBURY.

Mildenhall,
Dec, 7th, 1854.

My Dear Emily,

I am delighted to learn from your letter to
Fanny that you are really better, and able to enjoy
your dear Abergwynant, and that your journey
thither has answered its purpose. It certainly is a
delight of a place, and even at this season, I
suppose, has still much beauty; indeed, its luxuriant
Mosses would perhaps show to more advantage at
this season than any other; and the fine weather (if
you have had as much as we have lately) must have
been very enjoyable.

I was at Bury yesterday (I went in to visit the
gaol), and heard from Scott that a telegraphic
message had been sent down to Bury the day before
announcing Hanmer's arrival (in company with a
Dr. Martin) in London, and requesting Sally to go
to him; and that Norman telegraphed back *where*
my father and Sally were. Poor fellow, I hope very
soon to hear in what state he has returned. At any
rate, I am very glad he *has* come back, for it was sad
to think of the chance of his dying alone and uncared
for, far from his friends. I would hope for the best,
and pray that he may yet live to be a comfort to his
wife and his father.

1854.   The more one reads about the battle of Inkerman, the more thankful I feel that our dear Henry came off safe. It is not wonderful that the Russians should have been confident of success; their attack seems to have been well-planned, and all the chances must have appeared in their favour, especially after they had succeeded in gaining, unobserved, the position of attack. Nothing but the most heroic bravery could have saved our army from defeat. No wonder the English and French troops should be full of enthusiasm for one another. Even Kossuth, I see, much as he dislikes England, owns that the bravery of our army was astonishing. I am afraid that the end of the siege is still far off, and that our brave fellows will suffer much in the winter, though perhaps some of the newspaper accounts of the climate may be exaggerated; but it seems difficult to get at accurate information, even as to the climate we have to contend with. It is clear that the *extreme* south coast of the Crimea, under that wall of mountains, where the Arbutus and Andrachne grow wild, and the olive is cultivated with success, must have a mild climate; but I fear it may be very different on the exposed hills about Sebastopol. How lamentable is the loss of so many vessels on the 14th! and I am much afraid these may not be the last disasters of that kind we shall hear of.—I must own I extremely dislike the idea of this treaty with Austria; the crafty rascals at Vienna will humbug our soft, easy Ministers, and persuade them into a peace which will be advantageous to nobody but Austria; the blood of our brave soldiers will have been shed in vain, we shall remain *beffati*, with nothing but the 1854. barren glory of two or three bloody battles, with the satisfaction of having sacrificed thousands of valuable lives to no purpose, and having the battle to fight over again another time, perhaps without the aid of France. Or, if peace is not made in a hurry, perhaps we shall have to assist Austria against her own cruelly-oppressed subjects.

Pray give my love to my father, and believe me,

Your very affectionate,

C. J. F. BUNBURY.

Mildenhall,
December 31st. 1854.

My Dear Katharine,

In the present state of my knowledge of Ferns, I am inclined to think that the soundest principle is that indicated by R. Brown, in the " Plantæ Javanicæ,"—that modifications of the venation should be used for generic characters *when connected with the position of the sori;* —and having due regard also to habit. As an instance of what I mean :—Goniopteris is distinguished from Poly-podium, by a particular arrangement of the veins ; a nearly similar arrangement exists in Nephrodium as limited by Presl, while his Lastrea has the venation of Polypodium, but Goniopteris differs also from Polypodium in the position of the *sori* on the veins, in which respect it resembles Meniscium ; whereas Nephrodium and Lastrea agree in the position of the *sori*. Therefore I am inclined to think Goniopteris a valid genus, and on the contrary

1854. to keep Lastrea as only a sub-genus of Nephrodium, this last being the older name. Lastrea Thelypteris, however, in which there is a striking difference between the barren and fertile fronds, may possibly deserve to be distinguished, and in that case it should keep the name of Lastrea. You will find Presl's arrangement given, in a table at the end of the Genera Filicum, which I think you have; but I do not like his system so well as Mr. Smith's. It is a thousand pities Sir William Hooker does not go on with the Species Filicum. Matonia is one of the rarest of all Ferns; till lately, only one specimen of it was known to exist in Europe; so it will be a great prize for your herbarium.

It really makes one's heart sick to read of the condition of our army in the Crimea. That one of the finest armies that ever left England should be suffered to perish,—to rot, to fall to pieces in that miserable way, destroyed not by the enemy, but by exposure, privations, disease, famine; perishing with hunger and cold, when there are provisions and supplies in the harbour so few miles from them! And the Government newspapers would have us believe that it is (like a railway accident) nobody's fault! I do believe that the *least* share of blame, if any, ought to fall on the Goverement at home; but it is evident that there must be deplorable deficiencies, either of knowledge or care, in some of the departments *out there*. My father said, long ago, that "the English understand nothing of war, except the fighting"; and every day confirms the truth of this.

A happy new year to you, dear Katharine, and

many of them, and pray give my love and best wishes 1854.
to Harry and your children, and all our friends in
Harley Street and Queen's Road West.

Ever your affectionate Brother,

C. J. F. BUNBURY.

STUDIES.

(December). Of books *not* connected with natural
history, the principal I have read are the following :—

1. " Southey's Life." Six vols. One of the
best and most agreeable biographies I have ever
read'; his letters are admirable.

2. My Father's " Narrative of Military Events
from 1799 to 1810." Very interesting; but only a
part of it (the general introduction and the Egyptian
Campaign) was altogether new to me.

3. " The History of Sir Charles Napier's Ad-
ministration of Scinde," by Sir William Napier;
full of interest and instruction, though many
objections may be made to the style and taste of
various parts of it. The real greatness of Sir
Charles's character, the grandeur and comprehen-
siveness of his views, and the vigour and ability with
which he carried them out, are powerfully shown.
The most striking part of the History, is his
wonderful campaign against the Robber tribes of
the Cutchee Hills.

4. " Sir Charles Napier on the Misgovernment
of India," edited, after his death, by his Brother. An

1854. unsatisfactory book on the whole, though many things
in it are excellent; many sagacious remarks and
important suggestions, some brilliant bits of narrative,
and some quaint touches of characteristic oddity. It
might have had a more judicious editor.

5. "Memoir of Kossuth," by Susan Horner.
Gives a very clear and satisfactory view of the origin
and reasons and nature of the Hungarian revolution,
of the grievances of that people, and the misconduct of
the Austrian Government. As a biography I do
not think it successful; it fails in giving (what
Plutarch is so excellent in) a clear and vivid idea of
the man.

6. "Brace's Hungary." This I had read before
in '52, at Edinburgh, but I read it again as an
illustration of the History of Kossuth. It is quaintly
written, but gives one a very lively impression both
of the people and the country.

7. "Moltke's Account of the War between the
Russians and Turks in 1828-29." Dry, but
instructive.

I have also read with Fanny in the evenings,
lately, the 12th, 13th, 14th and 15th cantos of
Ariosto; and moreover, something of a very different
character, an important article (of seventy pages) in
the last *Edinburgh Review*, on "Prison Discipline
and the Reformation of Criminals."

In German, I have read about 120 pages of Von
Humboldt's "Ansichten der Nature," and I think
I am really beginning to make a little way in that
language.

Of Greek, I have read nothing, and of Latin very

little, but have very lately begun to read " Livy's 1855. Narrative of the Second Punic War."

On the whole, considering how much time is consumed in reading the newspapers (for one cannot avoid reading all that relates to this war), and in writing letters, and how small a part of each day I have really free from interruptions, I am not entirely dissatisfied with these seven months, though I certainly might have done more. I have spent more time out of doors, especially in looking after the plantations, than I used to do ; much to the benefit of my bodily health, with some improvement in practical knowledge, and not entirely without addition to my knowledge of the natural history of this district.

Mildenhall,
January 7th. 1855.

My Dear Katharine,

I must thank you again, and very heartily, for the really splendid collection of Assam Ferns which you have so kindly sent me. Many of the specimens, I see, are from the Khasya hills: a name now familiar to me from Joseph Hooker's interesting account of those hills,—of their peculiar climate, striking scenery, and rich vegetation. They have a greater allowance of rain, it appears, than any other place in the world where the fall of rain has yet been measured; so it is not wonderful they are rich in Ferns. I see among the lot my old friend Lycopodium cernuum, one of the plants I gathered in my very first walk at Rio de Janeiro. One of the Lygodiums, too, is extremely like the Brazilian L.

1855. volubile; and, what is much more extraordinary, the Osmunda looks to me exactly like the North American O. interrupta.

Our shrubbery is gay with the yellow Aconites, but I have not yet seen any Snowdrops.

Pray give our love to Harry and your children, and believe me,

Ever your affectionate Brother,

C. J. F. BUNBURY.

Mildenhall,

January 24th, 1855.

My Dear Emily,

I am much disappointed about the sad business in the Crimea: the ruin of our noble army through such scandalous mismanagement, the loss and disgrace to our country, the misery brought on so many thousands of families—altogether, it is too shocking to think of, even if our dear Henry were not involved in the danger. It does seem extraordinary that there should be such a total want of administrative talent throughout every department connected with our army. I remember my father said, at the very beginning of this campaign, the English understand nothing of war, except the fighting, and most fully has this saying been verified.

But I hope—no, I wish that a strict enquiry may be made into the causes of this disaster, and a signal example made of those whose imbecility or obstinacy has been the cause of it. But I expect no good from the present House of Commons, which seems to be influenced by nothing but party spirit.

It is delightful to hear such good accounts of Cissy 1855.
and the *infant Colonel*; he must not, however,
supersede dear little Emily. We hope to go to
London about the middle of February. Hanmer
and Sally, with some of the children, are coming to
us the end of this month, to stay a few days before
they remove to Bath.

Pray give my best love to my father; and with
Fanny's love,

Believe me ever,

Very affectionately yours,

C. J. F. BUNBURY.

Mildenhall,

January 28th, 55.

My Dear Katharine,

Certainly the state of public affairs is not
calculated to cheer one. On every side, clouds and
thick darkness. But God is over all; and one must
believe that public as well as private misfortunes all
tend to good. The misfortunes of our army in the
Crimea weigh much upon my mind. The destruction
of that army, which seems all but inevitable, appears
to me the greatest disaster, and the greatest disgrace
that has befallen our country in modern times.
What was the loss of Minorca—which put all
England into a frenzy a hundred years ago? What
was the destruction of Braddock's Corps? What was
the Caubul disaster, compared with the loss of such
a noble army? Destroyed, too, as it is, not by the
overwhelming force or superior courage of the enemy,
but by neglect and mismanagement. Well might

1855. Lord John Russell feel that there was no resisting a motion for enquiry under such circumstances. If the advice, which it now appears he gave in November, had been adopted, and Lord Palmerston made War Minister, with ample powers, matters might have been much better managed: but *now*, even if the change were to be made, I fear it would be *too late*, like everything else these Ministers have done in relation to the war. However, after all, one must not lay the whole blame on the Ministers; a great deal of it is due to the miserable system, the constant prevalence of interest over merit, and the tyranny of official etiquette and routine. The whole thing makes one sick at heart. God knows whether we shall ever see my brother again. If any of those now out there (except the General and his staff) come home alive, they will have cause to think it a most special mercy. I suppose Lord Raglan will come home some fine day, and report that he has buried his army.

I am busy reading in old Livy, the " History of the Second Punic War. He is a splendid writer. I was very much entertained by " The Life of Sydney Smith," especially his letters.

With much love to Harry and your dear little boys, and to all friends at Queen's Road West,

Believe me ever your affectionate Brother,

C. J. F. BUNBURY.

Mildenhall,
February 1st, 1855,

My Dear Emily,

I heartily concur in all your sentiments about the War and the Ministry, and I rejoice in the majority on Roebuck's motion, because it will, at any rate, relieve us from Lord Aberdeen, and because it shows that the nation is in earnest on the subject, and that, whoever may be Minister, must be in earnest too. So great a majority in the House of Commons that has hitherto been so staunch to the Aberdeen Ministry, is a strong indication of how thoroughly the general feeling of England is roused. I see in the paper to-day, it is reported that the Queen has sent for Lord Derby. I shall be very glad of this *if* it ends in placing Lord Ellenborough at the Head of the War Department. I conceive there is no man in England so fit for that office; but will he act with Lord Derby, or with anybody? We shall see. I hope the evils have not gone too far to be remedied. At any rate, I take it, no good can be done unless the War Minister has the powers of a Dictator, supreme over all boards and departments; power, as well as strength of will to over-rule all the impediments of routine and official etiquette and jealousy, and to remove without mercy or delay all incompetent and worn-out officials. My old friend Stafford's speech, was very instructive and interesting, and I have no doubt his account may be thoroughly relied on. What a picture it is! and what a humiliating contrast between our management and that of the French.

1855. Stafford, by his assiduous attention to the hospitals, has by all accounts, been doing a great deal of good. It is evident from his account, as well as others, that all the officials connected with the hospitals (and doubtless it is the same with the other departments) are paralyzed by routine and etiquette, and the extreme dread of responsibility; but Sidney Osborne says, moreover, that he has named to the Duke of Newcastle, several of them who are culpably negligent and inefficient.

Fanny had a letter this morning from Henry, dated the 15th January, quite cheerful as to himself, but giving a woful account of the army. He says, indeed, that the condition of the soldiers as to clothing, is rather improved, but that still, no *beggar* in England is so ill off as they are. He says that nothing is doing, and nothing can be done but to try to keep themselves alive till the Spring; that there is no sort of use in sending out any number of boy-recruits, who only serve to crowd the hospitals; and he wishes for a good number of stout Germans who have had some experience of war, and is therefore in favour of the Foreign Enlistment Act. But I hear of no results from that measure. I should like very much to know what my father thinks of it.

Poor Sir Thomas Cullum's death was very unexpected,—quite startling; for there was no man of my acquaintance, I thought, more likely to live to a hundred. Lady Cullum will no doubt feel his loss much, but I hope her worldly position will not be changed for the worse.

Have you seen "Sidney Smith's Life?" It is 1855.
not published, but Mr. Horner lent us his copy, and
we have been very much entertained by it. The
letters are capital. There was a remarkable union
of sturdy vigorous good sense with wit and humour
in Sidney Smith, and strong principles too, and
staunch courageous adherence to whatever he
thought right. Of poetry or romance, there does
not seem to have been a particle in his composition.
This is the only *new* book I have seen for a long
time. I am deep in the History of the Second
Punic War, in old Titus Livius ("*Tite Live*") : a
most interesting writer he is.

I was very glad to hear from Lyell that we are at
last really to have some more volumes of Macaulay
next autumn ; he (Macaulay) was willing to have
them ready by June, but the publisher thought it
best to wait till the autumn. I shall have to read
over again the greatest part of the first two, to
refresh my memory.

Pray give my love to my Father and to Cecilia
and her piccanninies.

Ever your very affectionate Stepson,
C. J. F. BUNBURY.

Mildenhall,
February 4th, 1855.

My Dear Mr. Horner,

I thank you very much for your kind note,
and your very agreeable present of poor Forbes's
writings. What I have yet read of them is interesting and very pleasantly written, and I do not doubt

1855. I shall find much entertainment and a good deal of instruction in the little book.   The portrait at the beginning is very characteristic.   Poor Forbes, I can never think of him without regret.

I am very happy to hear that you have been able to work up your Egyptian researches so far as to bring them before the Royal Society, and I hope to see them by and by in print ; they cannot fail to be important.

My dear, good Fanny has just made me a noble birth-day present of Murchison's new book of Siluria, which I am eager to read.   It looks, as one cuts open the pages, extremely tempting.

I am looking forward with very great pleasure to our visit to you, which I am sure will do us both much good, to me mentally, and to Fanny every way.

What a mess public affairs are in! our army perishing in the Crimea, and at home no one able to form a Ministry ! what will be the end of it ?   I am afraid at any rate we have little chance of a strong and united Ministry ; and yet there is nothing more essential at present.   We English are apt to boast of our practical ability, and of being a practical people ; what a comment upon that notion is the management of the present war.

With Fanny's best love, believe me,

Ever your affectionate Son-in-law,

C. J. F. BUNBURY.

To Lady Bunbury.   1855.

My Dear Emily,

We arrived here late on Friday evening. The railway is an immense advantage in this country, where there is scarcely anything to see except in the towns, and where the nature of the soil makes the formation of good roads difficult. All on this side of Cologne was new to me. We were very much pleased with Brunswick, where we arrived early enough on Thursday to have a good survey of it before night. It is a most quaint, picturesque, interesting old town, full of odd old houses with carved wooden fronts and fantastic gables, and odd projections of all sorts: just such a town as one sees in old German pictures, or in Retzsch's drawings. In the crypt under the Cathedral, we saw the coffins of the Brunswick Princes, in particular of "Brunswick's fated Chieftain," who was killed at Quatre Bras, and of his father, who fell at Jena. The Cathedral of Magdeburg is a noble building, and there is in it a bronze monument of an Archbishop, with numerous figures by Peter Fischer, of Nuremberg, that is perfectly admirable.

The country is certainly, for the most part, uninteresting, yet, near Minden, there is a pretty wooded chain of hills, and as we approached Brunswick we had a distinct, though distant view of the Hartz mountains, on which much snow still remains. To me, there is a great pleasure in

1855. travelling through a *historical* country, where so many names are associated with events and persons that one has read about. Minden, Hanover, Brunswick, Magdeburg—the Weser and the Elbe, are names that immediately call up interesting recollections.

(May 11th). I am passing my time here very pleasantly indeed, and have so much to do and to see, that I really can with difficulty find time for letter-writing; but above all, I must tell you of our two interviews with Humboldt, which I consider great events. I brought a letter to him from Dr. Hooker, and he is, besides, acquainted with Mr. Horner and the Lyells. We called upon him by appointment, on the 7th, and he received us most courteously; but we were interrupted by some other visitors coming in, and could not profit by his conversation so much as I had hoped. However, yesterday he returned our visit, and sat with us an hour, talking the whole time most agreeably. He is a delightful old man, with all the courtesy and polish of an old Frenchman, and with a vivacity and activity of mind that are perfectly wonderful in a man of 85. He is a little bent, but still hale and fresh looking, and so strong that he walked hither from his own house, in a rather distant part of Berlin, and back. He has all the volubility of speech that I have often heard of, but you may well suppose I was right willing to listen, and did not wish to say much. His conversation is most interesting, and what is particularly striking is his eager interest in all that is going on in all the world of science, his acquaintance with all the newest researches, and his constant desire for fresh information.

As Fanny says, he is a remarkable contrast, in this 1855. respect, to Mr. Rogers, who, for many years, has never cared to talk of anything but the past. Humboldt does not appear at all egotistical, though he is easily led to talk of the countries in which he has travelled (and I particularly wished to hear him on that subject); he did not dwell on his own adventures or writings. Desirous as I had been to see him, from my great delight in his writings, I have not been in the least disappointed, but rather have had my expectations surpassed; and certainly I looked upon him with more interest and veneration than I should have felt for the assembled Sovereigns of Europe. We are very well situated here, directly opposite Dr. Pertz's house, where we dine and spend every evening, and we are quite in the centre of all that is to be seen.

This central part of Berlin including the Library, the University, the Museum, the Arsenal, the Theatre, and the King's Palace, is certainly very handsome: and there is so much to see, that the fortnight we originally settled to remain here will be a very scanty allowance. I have been particularly struck with Rauch's monument of Frederick the Great, and Kaulbach's frescoes in the new Museum. The gallery of pictures, though it does not contain many absolutely first-rate works, is exceedingly interesting and instructive, from the completeness of the series of works of all the schools and its admirable arrangement. The Library which Dr. Pertz has shown to us, is a magnificent collection, and likewise in most perfect order. — Besides

1855. Humboldt, I have seen most of the eminent scientific and literary men of Berlin, and made acquaintance with several of them, from whom I have received great civilities and attention ; and I have learned much from the Botanic Garden and the collections of the University. I need not say that we receive every kindness from Dr. Pertz. His three sons are very amiable and pleasing young men. Altogether I shall be quite sorry to leave Berlin, and would gladly remain a month here, but we want to see Dresden and Leipzig, and Eisenach, and to be back in England in time to see Henry, before he has to return to the army.

I hope you and my father are in tolerable health, and that the weather at Barton allows you to get about in the garden. Here the weather is tolerable, and that is all I can say. It is certainly not a poet's May. Fanny has heard from Cecilia since we have been here, and a nice warm-hearted letter from Harry* Poor dear Cecilia; I feel very much for her, as well as for Emily;† and what numbers more there must be of wives in England suffering the same sorrow! I trust, however, from her account, that Henry's health is really improved.

Pray, give my love to my father, and believe me,

Your affectionate Step-son,

C. J. F. BUNBURY.

---

* The son of his brother Hanmer.
† The wife of William Napier (afterwards General), Sir George Napier's third son.

JOURNAL.

Berlin. Visited Professor Alexander Braun, who was formerly at Carlsruhe, and is now head of the Botanical Department here.

———————

I have had the great pleasure to day of visiting Baron von Humboldt. He received us with the utmost courtesy. He is a very fine old man, and wonderfully young for his age, which I believe is 85 ; his stature shorter than I had supposed, and he is a little bent, but quick in his movements ; his countenance cheerful, and his manners peculiarly courteous and pleasing. He gives me the idea of a Frenchman of the old school, with their characteristic union of vivacity and cheerfulness. He was very ready to talk, and his remarkable volubility was what I had expected from previous accounts of him.

Unfortunately, we were interrupted by some ladies coming in, so that he did not talk, as I had hoped he would, of his own travels. He said he had lately heard from Bonpland, who had been making extensive voyages on the rivers connected with the Plata. He talked of the Mexican volcanos ; remarked that the line of volcanic vents there, is at right angles with the general direction of the mountain

1854. chain; and said that he had received specimens of
the lavas of Popocatepetl and Orizaba, which (as I
understood him) are identical in composition with
those of Chimborazo.

The variety of felspar occurring in these lavas is
that called *oligoklase*, whereas the lavas of Etna (as
I understood) contained the variety called *labradorite*.
—Humboldt showed us a medal struck in the newly-
discovered metal *aluminium*, the metallic basis of
alumina.   It is of the colour of platina, or of rather
dingy silver, and as light as glass, and the impression
was sharp and good.   He mentioned that he is now
engaged in preparing for the press the 4th and last
volume of "Kosmos," which is entirely geological.

(May 9th.)   Professor Beyrich showed me the
collection of fossils of the University, which is a
very fine one.

(May 10th.)   A memorable day for us.   We had
a delightful visit from M. de Humboldt, who stayed
with us fully an hour, talking most agreeably the
whole time.   His conversation is rich, varied, lively,
and instructive, like his writings; and it is marvellous
to see, at his great age, the activity of his mind, his
eager interest in all that is going on in science, and
his unflagging desire for fresh information.   He had
looked over my paper on Buenos Ayres, since our
visit to him on the 7th; enquired much about the
botanical relations of that country, and was particularly desirous to know what species it has in common
with other Continents.   He told us that Bonpland
discovered the Victoria Regia, in the river Paraguay,
some years before its discovery by Schomburgh in

the Essequibo, and even sent seeds of it to Paris, 1855. but they did not germinate. It is called in that country *Maïs de agua*, and its seeds are commonly eaten. He said that he had received drawings of the gigantic tree from the West Coast of North America, which Lindley has called Wellingtonia, and that one of these trees was said to be, by actual measurement, 420 feet high; but, as he remarked, "*c'est un peu fort.*" He does not think the genus distinct from Cupressus. Talking of the variability of Coniferæ, he mentioned that he had seen the common Pine of the mountains of Mexico, which has normally five leaves, varying with four, and even three.

He remarked the singular fact of the occurrence of true Pines—the Pinus occidentalis, Swartz, on the Isla de Pinos, near Cuba, which is "*à fleur d'eau,*" whereas, in Mexico, there are no Pines below the elevation of 5,000 or 6,000 feet. He said that Columbus had remarked in his Journal, as a singular fact, the association of Pines with Palms—*Pineta et Palmeta*—either in Cuba or Hayti (I am not sure which), and that he, Humboldt, had seen the same thing in Mexico. Speaking of Endlicher's synopsis of the Coniferæ, he said that Endlicher had latterly given up botany for classical and philological studies; had paid much attention to the Chinese language, and had even written a Chinese grammar, "*qui ne valait pas grande chose*"; but that he had also speculated, very unfortunately, in railways, and his misfortunes in this way led him to destroy himself.—Speaking of travelling in tropical countries, Humboldt gave us a

1855. most lively account of the contrivances to which he and Bonpland had recourse, to procure a respite from the torment of mosquitos on the Orinoco; of the suffocating ovens into which they crept to dry their plants, after the insects had been driven out by wood smoke; and of their climbing up to sleep in little huts, or lodges, elevated on the tops of tall posts, to be above the *stratum* of mosquitos.   He said that in some of the missions, they saw people burying them-selves under some inches of soil, in order to sleep, leaving their heads only out, with their faces covered by a handkerchief, and collecting the cattle round them, to divert the insects from themselves.—He mentioned the remarkable difference between the table-land of Mexico and that of Quito: that the Mexican plateau is intersected by no remarkable ravines, but is so level that a coach and six might be driven along it for many degrees of latitude North of the City of Mexico; whereas the table-land of Quito is cut by ravines of extraordinary depth and abrupt-ness, some of them even 11,000 or 12,000 feet deep, so that often it is a laborious day's work to descend to the bottom of one of them; and the climate at the bottom is intensely tropical, while it is cold on the plateau above.   Fevers prevail much in these deep valleys.

Pertz tells me that Humboldt usually stays at the Palace till eleven o'clock at night, then goes home and begins writing, and often does not go to bed till three in the morning.   He is a great favourite with the King, who yet never consults him on political matters, as he (H.) disapproves strongly of the course

at present pursued. After dinner, Pertz took me to 1855.
the Academy, where I saw several eminent men:
Ehrenberg, Encke, Rose, Mitscherlich, Ritter the
geographer, the two brothers Grimm (Jacob Grimm
has a fine poetical head), and others. In the evening,
a small party at the Pertzes', where I had some
conversation with Dr. Ewald, the geologist, Professor
Gerhardt, and especially with Dr. Peters, a very
clever young man, who made a most arduous and
hazardous expedition in Eastern Africa, exploring up
the river Zambesi, from the Mozambique coast. He
seems to have enjoyed this adventurous journey very
much, and says that his greatest wish is to go
back to those countries. The climate is, however,
extremely unhealthy. He brought home a large
zoological collection, and was so fortunate as to lose
only one chest out of fifty-two that he had filled with
specimens, although all his baggage had to be carried
by men. He mentioned a remarkable fact which
often struck him in his travels: that where the
vegetation was most luxuriant, animals were least
abundant. He often travelled a considerable way
through thick forests, without seeing or hearing even
a bird,—very unlike the state of things in South
America. In that part of Africa, it seems, there are
no Apes, like the Chimpanzee of the West Coast;
the only animal of the monkey family that Dr.
Peters met with, was the large Baboon, or Cyno-
cephalus, the same that is found in Abyssinia.
Lions are very numerous.

(May 12th). Professor Braun took me to the fine

1855. garden of Monsieur Decker, which is particularly rich in Tree Ferns. He has a relation, Monsieur Karsten, at Caracas, who sends him the finest plants from that country: and his gardener, Monsieur Reineke, seems to be particularly successful in cultivating them. The Tree Ferns are the finest and most numerous I have seen in any collection: one in particular, an Alsophila obtusa, from Puerto Cabello, is by far the finest specimen of an arborescent Fern that I have seen in Europe, and really gives one a just idea of their beauty in tropical countries. None at Kew are comparable to it. The trunk is *eleven* feet high to the base of the crown of leaves.

May 13th.

Went with Pertz, his son Hermann, Joanna Horner, and a large party to Rüdersdorf, a place a few miles nearly East of Berlin, to examine the Muschelkalk formation which is well-developed there. The village of Rüdersdorf lies in a small valley. The Rüdersdorf quarries are very important in an economical point of view, as being the only good building-stone in the vast sandy diluvial plains of Northern Germany. They are of very great extent, having supplied most of the building-stone for Berlin, and much for other cities: and they are known (as Pertz tells me) to have been worked in the 13th century.

May 14th.

Visited M. de Humboldt again, and found him

JOURNAL. **333**

courteous and agreeable as before. Monsieur Pictet 1855. of Geneva, with his son, came in and stayed some time. Humboldt talked much of astronomy and meteorology, on which subjects I could not always follow him. He spoke with great admiration and affection of Arago, of whom a fine bust was in the room. He thought it important that there should be *separate* observatories for astronomy and meteorology, and that the latter science would not till this was done, make the progress it might do. He urged me to visit Rauch's studio, and highly extolled the last work of that great sculptor, a marble group of Moses with his hands upheld in prayer by his two attendants; the subject, he said, was suggested by the king, and the combination of the three figures in one group, might be compared to the Laocoon. Thence he went on to remark, that the story of Laocoon was evidently an Indian myth; that the idea of the gigantic serpents, destroying men in their coils, would never have originated in Greece or Asia Minor, but must have migrated from India. Then he spoke of the great serpents, species of Boa, which he had seen in his voyage on the labyrinth of rivers connected with the Orinoco; that while passing in their canoe through the inundated forests, he and Bonpland had seen many of these great snakes, 10 or 12 feet long, swimming with their heads raised above the water. In the same streams (as he has recorded in his travels) were numerous dolphins or porpoises, leaping and gambolling like those of the sea; and the little monkeys, which he kept alive in his boat, where much frightened at the

1855. noise made by these dolphins.   This is a good
specimen of the discursive style of his conversation.
He said that he had been attacked by a sort of
scorbutic complaint, which had for some time almost
crippled one of his arms, in consequence of the
hardships of his expedition on the Orinoco, and
especially the excessive damp to which they were
exposed night and day ; the quantity of decaying
vegetable matter being often so great, that a phos-
phorescent light was diffused all around.   He
showed me a new set of maps of Isothermal Lines
by Dove, and a large map of the Polar regions, with
all the newest discoveries.

May 15th.

Saw a part of Willdenow's Herbarium, which,
together with the rest of the royal Herbaria, is under
the care of Dr. Klotzsch, and kept in a rather
handsome building opposite to the further end of
the Botanic Garden, at Schöneberg, about two
miles from Berlin on the Potsdam road.

May 16th.

I visited Dr. Ritter, the celebrated geographer, a
fine old man, very conversible.   He told me many
things about Dr. Barth's travels in the interior of
Africa.

We drank tea with Count von Beust,* who has had
the charge of the mines of the Prussian dominions.
He has travelled much, and gave us much information about Spain, in which country he made an

* The Uncle of the Minister.

extensive tour some years ago. He says that the 1855. silver mines in Spain (near Guadalaxara, I think), which have not been opened many years, are very rich ; the lead mines not so productive as formerly. The quicksilver mines of Almaden are still very rich, but the demand for mercury is not quite so great as it used to be, since a method has been discovered, and brought into use in the Saxon mines, of separating silver from its ores without amalgamation.

We had another visit from Von Humboldt, but a short one. He was very courteous and pleasant. He expressed great satisfaction at Joseph Hooker's appointment at Kew. He told us that the Dutch government are trying to cultivate the true Cinchona in the Island of Java ; a well-qualified person was employed to procure plants from the neighbourhood of Loxa, and plantations were formed on the mountains of Java, at the height of 6 or 7,000 feet. He observed, however, that as the Cinchonas are not naturally social plants, but grow scattered amidst more robust trees, of which they seem to require the shelter, there may be some doubt whether the attempt to cultivate them *by themselves* would answer. He spoke of the imperfect success which has attended the cultivation of Tea out of China.

Professor Lichtenstein showed us part of the collection of birds in the University Museum, which

1855. is very extensive, and appears very rich, especially in series of specimens showing the variations of each species, according to sex, age, and other circumstances; the specimens appear also to be carefully labelled, and the collection to be a very instructive one. Dr. Lichtenstein, Professor of Zoology, is the same whose travels in Caffraria, at the beginning of this century, are well-known. He is a lively and pleasant old man.

LETTER.

Hotel des Princes,
Berlin,
May 20th. 1855

My Dear Father,

Of men not scientific, the most eminent I have seen is Ranke, the Historian; a very odd man he is, scarcely agreeable, but his conversation interested me. Among other things, he remarked that all our best histories of England end just when the history begins to be most interesting to other nations; and, speaking of Macaulay, he said (I think very justly) that Macaulay has introduced King William so magnificently, and begun with such a highly-coloured portrait of him, that he will not be able to keep him up to the same elevation. Ranke's face has a singular expression of shrewdness, almost of cunning, rather than power.--I have seen, and merely *seen*, the great sculptor Rauch; a very fine-looking man, but seemingly difficult to be acquainted with.

His statues are admirable, to my thinking. Kaul-

bach's frescoes are certainly very fine, his power of 1855.
drawing is wonderful, and his colouring agreeable;
but in his large compositions, to my thinking, there is
too much *enigma*, too many deep and recondite mean-
ings,––and sometimes a rather bewildering intermixture of human and supernatural agents.––The
galleries of the New Museum, in which the collection of
casts from sculpture is arranged, are most beautiful;
the gallery of pictures extremely interesting and instructive, being, I believe, the most complete and
best arranged *series* of works of all the schools, and
all the principal artists of Italy, Germany, and
Flanders, that is to be seen anywhere. It is particularly rich in early German and Flemish works.
There is one of the finest works of Van Eyck (the
wings of the great picture which is at Ghent); some
singularly characteristic and striking pictures by
Lucas Cranach; and the very finest Rembrandt I
ever saw––the Duke of Gueldres threatening his
imprisoned father. Correggio's Leda (formerly in
the Orleans Gallery), is a most lovely picture.

I am very much pleased with our brother-in-law,
Chevr. Pertz, and I hope you will one day know him,
for I am pretty sure you would like him. He is a
quiet, moderate, reasonable man, very free from
prejudice, I think; cheerful and conversible, and
seems thoroughly conversant with the modern history
of Europe, as well as with that of the middle ages.
He is now preparing to write the life of Gneisenau.
He is in a most enviable position here, in charge of
this splendid library. His three sons are very
pleasing young men, especially Hermann, the

1855. youngest.   Altogether, I have passed a most agreeable fortnight at Berlin, and shall always remember
it with pleasure, although the weather has not been
genial.   We set off, the day after to-morrow, for
Breslau, where I want to see a Professor famous in
fossil botany; thence to Dresden, and then we come
back to Berlin, to spend some more days here before
we turn our faces homewards.

I am very glad indeed to hear that Henry's leave
is extended to the middle of July.   I trust this will
allow time for his health to be really re-established.

Believe me ever,

Your affectionate Son,

C. J. F. BUNBURY.

JOURNAL.

Breslau,

May 23rd.

Visited Professor Göppert, who was very obliging,
and showed us several interesting specimens.   He
lives in a pleasant house almost outside the town,
near the Cathedral.

May 24th.

Visited, at the University, Dr. Ferdinand Roemer,
whom I saw in '48 at Bonn, and who has lately come
to take a Professorship here, and has the charge of
the collections of mineralogy and geology in the
University.   He is a very lively and very clever
man, speaks English well, and seems well acquainted
with what is doing in the natural sciences among us.

He showed us a few specimens of fossil plants,—not 1855. of great importance,—and some fine minerals; in particular, a most beautiful group of crystals of *red silver-ore* (the finest, he says, he ever saw), and a splendid crystal of apatite, of a bluish green colour, with both terminations perfect—both these from Saxony. Several good specimens of ærolites: among others, part of a mass of meteoric iron (with very little extraneous mixture), which fell through a cottage at Braunau, in Silesia, only a few years ago. He showed us the curious figures (results of crystalline structure) which are rendered visible on the surface of a piece of meteoric iron by washing it with dilute sulphuric acid.

Dr. Roemer is now one of the editors of the *Lethaea Geognostica.*

After dinner, we went with Professor Göppert to the Botanic Garden, which is close to his house; not of great extent, but prettily laid out, and apparently well attended to. The hot-houses are small, and, I should think, not well constructed, but contain many rare plants, though for the most part small specimens.

The Arboretum of this garden is good; the trees grouped according to families, and in part picturesquely arranged along the margin of a piece of water.

In one part of the Botanic Garden are arranged some specimens illustrative of remarkable peculiarities of growth. Among others, are specimens of that singular phenomenon in the growth of fir trees, which the German botanists call *Ueberwallung*, and which Göppert has particularly described. It appears

1855. that, when Fir trees grow densely together, in thick
woods, the roots of different trees of the same species
actually inosculate, and become grafted into one
another below the surface of the ground, so that
nourishment may be conveyed from one tree to
another, and even the stump of a cut-down tree
continues, for some time, to form fresh wood. This
happens chiefly in the Spruce and the Silver Fir, but
has been lately observed by the French in Pinus
maritima (Pinaster).—There are specimens also of
Fir trees with the base of the stem stilted up on
ærial roots, considerably above the surface of the
ground, as in Pandanus and Iriartea exorrhiza;
this happens when, in thick and damp forests, a Fir
germinates, and grows from a seed lodging on the
decaying prostrate trunk of another tree. As it
grows up, its roots shoot through the decaying
substance of the prostrate tree to the ground, and
so, when the other has quite decayed and fallen
away, they remain exposed to the air.—Here also is
a section of a common oak, with 300 rings of growth,
and another with 400; also a gigantic stump of the
extinct Pinites protolarix (from the Brown Coal), in
pieces, measuring altogether 40 feet round. Here
is a good collection of Ferns in the open ground; in
particular, Onoclea sensibilis (the barren fronds are
of a most beautiful delicate light green), Struthi-
opteris Germanica, Osmunda regalis, Aspidium
acrostichoides, and many others.

The beautiful public walks around Breslau are
under the care of Prof. Göppert. They are adorned
with a variety of trees, which are carefully labelled.

I observed some remarkably large trees of Acer 1855.
dasycarpon.

Along the margins of the water in the Botanic
Garden, we saw great numbers of the large green-
backed frog, Rana esculenta.

Dresden,

May 27th.

Professor Geinitz called on me very early, and
took me to see the collection of fossils, which is in a
part of the building called the Zwinger. The collection appears a very fine one, and in excellent order,
though all formed by Geinitz himself since 1849; for
the previously existing collection was all destroyed
by fire, together with the building, in the revolutionary
outbreak of that year. The present collection is
very rich in fine and instructive specimens of the
fossil plants of the Coal formations of Saxony, which
Professor Geinitz has illustrated in a most beautiful
work very lately published. Many of these specimens
are especially valuable, as showing the connection
between different portions of plants which have been
described under distinct names, though really belonging to the same thing.

May 28th.

After spending the morning in the picture
gallery, we drove out after dinner to Tharand,
nine or ten miles from Dresden in the direction
of Freiberg and the Erzgebirge. It is an extremely pleasant drive. First leaving the City,
we had a distant view of the curious abrupt
mountains of the Saxon Switzerland. Passing the

1855. village of Plauen, enter a very picturesque winding defile, between bold rocky hills, where dark masses of rock stand forth finely amidst thick wood; the pretty little river Weisseritz, bright and rapid, flowing along the bottom. The fresh bright green that now clothed all the trees, contributed to the beauty of the scene. The rocks are of a fine crystalline red syenite. Above this pass the valley opens out, and is very fertile and beautiful, gay with fruit trees in profuse blossom, and bounded by wooded hills. Higher up, another rocky pass; then again another beautiful wide basin.

The little town of Tharand very prettily situated, in a gorge between steep and beautifully wooded hills. We walked up one of the hills, to a church in a very commanding situation. Rain had just fallen, and had called forth amidst the foliage a number of very large snails,—I believe, Helix pomatia. We saw also a singular reptile, I believe a Salamander, in shape much like a water-newt, but with a very large thick head, and brilliantly variegated with large patches of deep saffron yellow on a jet black ground. The rocks here appeared to be of a coarse grauwacke slate.

LETTERS.

Hotel des Princes.
Berlin,
May 31st, 1855.

My Dear Lyell,

I have long been thinking of writing to you, but have really had great difficulty in finding time.

Now, having just returned from our Breslau and 1855.
Dresden tour, I have so much to tell, that I really
must *make* time for a letter. Our tour hitherto has
been in a high degree satisfactory and instructive,
as well as exceedingly pleasant. I have had most
especial pleasure in making the acquaintance of
Humboldt, who is indeed a delightful and admirable
old man, and whose conversation quite comes up to
the expectations I had formed. He has been most
courteous to us. Indeed I have met with all possible
attentions and kindness from the scientific men of
Berlin : but as you have been here so lately, I will
proceed at once to Breslau, which will have more
novelty for you. Göppert was extremely polite and
obliging, and gave me as much time as he could
spare from his academical duties, but he was very
much engaged with lectures and examinations ; he
speaks French very imperfectly indeed, and English
not at all, and is moreover very deaf, so that our
conversation was not quite as fluent as it might have
been, though Fanny was a very good interpreter.
I was rather disappointed with his collection, which
is in great disorder, except the specimens of Amber,
which are very numerous, and exceedingly curious
and interesting. You know his book on that subject,
and he showed me the original specimens there
figured, exhibiting the mode of occurrence of the
Amber in the wood of the Pinites succinifer, the
structure of the wood, and many other curious
details. But since the publication of that work, he
has got a great number of additional specimens,
which he is preparing to publish, and which he

1855. showed me, containing fragments of plants, which throw a most curious light on the contemporary Flora. Some of them are in a really wonderful state of preservation, especially the capsule of an Andromeda, which looks as if it might have been gathered yesterday. He affirms that *many* of the plants of the Amber formation are *specifically identical* with those of the present day; and certainly, as far as the specimens go, there seems to be no visible difference. He says there was at that period a remarkable mixture in these countries, of plants now characteristic of very distant countries, and even of very different climates; and so indeed it must have been, if the Libocedras Chilensis from the Southern Andes, the Thuia occidentalis of North America, and the Lapland Andromeda hyproides co-existed, as he concludes from his materials. Certainly the specimens do very closely resemble those plants, but it may be doubted whether such small fragments (scarcely any of them longer than one's thumb nail), and those, in the case of the Libocedras and Thuia, without fruit, are sufficient for the positive identification of species.—Göppert strenuously maintains that the *Stigmaria* is *not* a root, but a complete and independent plant, a floating water-plant: and he showed me several specimens which he considers conclusive on this point, as being complete and perfect individuals, entire at both ends, but I must own the specimens were not to me quite decisive. I suggested that different things might have been called Stigmaria; but Professor Göppert is positive that his plant is identical with that described by Dr. Hooker; and certainly it *has* the same structure and arrangement of vessels. However, I found afterwards at Dresden, that Professor Geinitz had taken the same view, as to the *duality* of Stigmaria, which in truth has occurred to me more than once. The Stigmaria *inæqualis* of Göppert, according to Geinitz, is the root of one or more species of Lepidodendron, particularly of Lepidodendron veltheimianum; and he showed me specimens from the Saxon coal mines closely corresponding with those sent by Mr. Brown, from Cape Breton. Now, it appears to me, that most, if not all of the North American specimens I have seen of Stigmaria, belong to this *inæqualis*. But the true original Stigmaria *ficoides* is, according to Geinitz, a distinct thing, and *this* he agrees with Göppert in considering as an independent, self-contained, self-sufficing vegetable. On this point, I must, for the present, suspend my opinion. The outward differences between the two are not very striking. What you have given at p. 371 of your Manual is certainly the *inæqualis*.

I was mightily pleased with Professor Geinitz and his collection. I have hardly ever seen a more beautiful or more instructive set of coal-plants than he has got together from the Saxon coal-field. The collection too has been all formed within the last five or six years, as the previous collection was almost entirely destroyed by fire, together with the building, in the revolutionary tumults, in which also the gallery of pictures narrowly escaped destruction.

1855.   The Professor was most kind and obliging; he
was just starting on a geological tour for the Whit-
suntide holiday, but he came to me before eight
o'clock in the morning, to take me to the museum,
showed it to me in a most agreeable manner, and
gave me a great deal of valuable information; and
on going away, he recommended me to Mr. Large,
the curator, so that I was enabled to visit the collection
again and again at my leisure.   Geinitz appears to
me a really clever man, his new book on the fossils
of the coal formation of Saxony (which I have
bought) is one of the most beautiful I have seen.
As for Göppert, all his writings that I have read
give me the idea of a very industrious, painstaking
and accurate man, but *not* naturally a very clever
one.   It would not be fair to say that my conversation with him gave me the same impression, for we
had so little language in common, that it would not
be fair to conclude anything therefrom. He certainly
is a very industrious and busy man, for besides his
various works on fossil botany, and his lectures and
examinations, he has the charge of the Botanic
Garden, which is a very pretty one, very nicely
arranged and rich in a number of species and in
rarities, though they are sadly cramped for room in the
hot-houses; and he has also the care of the beautiful
public walks which surround the city, and which are
adorned with a variety of fine trees and shrubs.

At Breslau we saw also Dr. Ferdinand Roemer,
who was formerly at Bonn, and whom I think you
know; he has very lately been appointed Professor of
Geology and Mineralogy at Breslau.   He also

received us very civilly, and talked much and 1855.
agreeably; he appears to me a very acute and
clever man. He is now one of the conductors of the
Lethaea Geognostica, in conjunction with Bronn.

Here, at Berlin, I have been exceedingly pleased
with what I have seen, particularly with the Botanic
Garden and Herbarium; and with Professors Braun,
Beyrich, and Lichtenstein, and Dr. Ewald. Joanna
will have told you of our capital geologizing expedition to Rüdersdorf, where I saw the Muschelkalk for
the first time. I shall say nothing of personal
adventures, nor of matters unscientific, because
Fanny has just sent off a long letter to Mary,
containing a full account of everything of the sort.
Suffice it to say, that I have been charmed with the
pictures at Dresden, and much interested by the
galleries here. It is quite unnecessary to say that
we enjoy the society of Leonora and George Pertz,
and of the young men, but we have missed Joanna
much since we came back. Humboldt was much
pleased with the extract from Mr. Prescott's letter,
relating to him, which Mary sent. I hope he will now
have received the 9th edition of the " Principles,"
but I have not yet seen him since our return. It is
a capital thing,—Joseph Hooker's appointment at
Kew; though the allowance is not magnificent, it
must be in many respects the most desirable situation
for him, and, of course, he will ultimately succeed
his father.

Much love to Mary, and to the party at Q. R. W.
Ever yours affectionately,
CHARLES J. F. BUNBURY.

JOURNAL.

1855. Professor Gustav Rose showed me the collection
of minerals in the University; it is a very fine one.
There is a fine set of specimens of meteoric iron and
ærolites; among them, models of the two masses of
meteoric iron which fell at Braunau, in Silesia, in
1847, and a piece of one of them showing a very
distinct crystalline cleavage, according to the planes
of the cube.   It is very nearly solid iron, with the
usual alloy of nickel, but with very little extraneous
mechanical admixture; but a small mass of sulphuret
of iron is imbedded in the midst of it.   Professor
Rose observed that these masses of *iron* have in very
few cases been *seen* to fall, so that this well-ascertained
instance is the more valuable.   The Tennessee mass
of iron was likewise seen to fall.

The ærolites or meteoric stones, of which the fall
has been observed, consist chiefly of stony matter,
but contain a greater or less quantity of grains of
metallic iron, alloyed with nickel.   The meteorite of
Juvenas, in France, however, Professor Rose told
me, contains *no* iron in a metallic form; it is a
granular compound of augite and a felspathic mineral
(anorthite?), which never occur so associated in any
known rock.   Here is the latest ærolite that has been
recorded; it fell as lately as September last (1854),
about six German miles from Berlin; the exterior
has the characteristic black, slaggy-looking crust, in

a very marked degree; the interior is whitish and
finely granular, but with disseminated grains of
metallic iron distinctly visible to the naked eye.
Professor Rose took much pains to explain to me the
characters of the different felspathic minerals: common or true felspar, albite, oligoclase, labradorite,
and anorthite. The last four, as I understand him,
nearly agree with one another in their crystalline
form, in which they essentially differ from common
felspar; but they are distinguished among themselves
by chemical composition, specific gravity, and mode
of occurrence. In common felspar, the principal
terminal planes of the crystals meet the lateral ones
at right angles; not so in the others. Common
felspar, like the others, is usually in the form of twin
crystals, but these are combined in a different manner,
and their terminations never exhibit re-entering
angles, as they do in albite and oligoclase. Common
felspar forms a constituent part of the *older* crystalline
rocks, granite, syenite, gneiss, porphyry; but never
occurs (so I understood the Professor) in lava or
basalt. The large crystals found in the granites of
the Riesengebirge, of Elba, and of Baveno, are true
felspar; so are the large imbedded crystals in por-
phyritic granites, and the fine red felspar in the
granite of Egypt. The opalescent felspar of Norway
is true felspar, not labradorite. The green felspar
("Amazon-stone") of Siberia is likewise true felspar.

Oligoclase very often occurs, together with common
felspar, in granite; often the two may be known by
their colour, the felspar being red, the oligoclase
white or yellowish, as in the beautiful granite of the

1855. great boulder at Fürstenwald, from which the magnificent basin in front of the Museum here was made. But sometimes *both* are whitish, and then the oligoclase may be known by the peculiar striæ along the lateral face of its crystals; these striæ, which depend on the peculiar mode of aggregation of its crystals, are never found in common felspar. *Albite*, which comes nearest to oligoclase in its characters, never occurs as a constituent part of rocks, but always crystallized in cavities. It contains more silex than oligoclase. *Labradorite* occurs chiefly in lavas and basalts; it characterizes particularly the lavas of Etna, in which it is the only felspathic mineral. It is a little heavier than oligoclase, and contains rather less silex. *Anorthite* seems to be a rare kind of felspar, hardly found except in the cavities of the old lavas of Somma, where it occurs in small crystals, in company with idocrase, garnet, nepheline, etc.

I saw here also the largest piece of amber that was ever found, valued at the time (1804) at 10,000 crowns. It is mentioned in Murray's Handbook. Also a specimen of amber imbedded in bog iron-ore, from some part of Prussia. Models of the two largest pieces of native gold found in the Russian dominions. Many specimens of platina from the Ural, and one particularly interesting, being mixed with chromate of iron. As chromate of iron has been found only in serpentine rocks, and there are mountains of serpentine at the sides of the valley in which the platina is found, this points to the probable original source (hitherto unknown) of the platina. Splendid

masses of malachite, from the mines of the Ural. A 1855.
very large topaz, of a beautiful pale transparent *blue*,
brought by Rose himself, and Humboldt, from their
Siberian journey ; and very fine beryls from the
same country.

Many curious pseudomorphous varieties of quartz,
(one of which in the form of Datolite) has been
described under the name of Haytorite.

I afterwards visited M. de Humboldt, who was, as
before, extremely communicative and agreeable. I
found that he had already run through my little
book on the Cape of Good Hope, which I sent to
him only two days before. He mentioned that he
had seen some Caffers who were brought to Berlin
in the winter, and had been struck with their
insensibility to cold ; they went about half naked in
a severe Berlin winter, when the thermometer was
down sometimes to 12 deg. below zero centigrade,
yet they did not appear to suffer from the temperature, nor did they *catch* cold. This was a great
contrast, he remarked, to the Negroes, who are
extremely chilly. He talked much of the geography
of plants, and touched, *inter alia*, upon Forbes's
essay on the origin of the Flora and Fauna of Great
Britain, which he thought rather too hypothetical.
He objected to the plan adopted by Schouw and
others, of naming the different regions of botanical
geography as the regions of such and such families of
plants, remarking that it was an attractive method,
but in many respects fallacious ; that there are indeed
some regions, such as Australia, which might be
characterized by special families of plants ; but that

1855. the plan could not be applied generally without producing false impressions. That there are moreover,
various ways in which the predominance of particular
families of plants in particular regions may be
understood ; either in reference to the large
proportional number of species of such families in
the Flora of a country, or to their being confined, or
nearly so to such country, or to the great extent of
surface occupied by social species, as the few species
of Heaths in Northern Europe. For himself, he
had always used in his works, the method of
numerical quotients (i. e. fractions expressing the
proportion of the number of species of each family
to the total number of species in the whole
Flora), and he thought it gave the truest idea of the
vegetation of a country. He mentioned to me some
maps of the botanical geography of Europe and of
Germany, which he had drawn up on this plan for
Berghaus's Physical Atlas. He spoke of the
singular fact in botanical geography observable in
Siberia, where without any change of level or
perceptible difference of climate, a small river forms
an absolute limit to the eastward range of several
very common European species.

In the evening, at a small party at the Pertz'.
I had much talk with Lepsius, the celebrated
Egyptian traveller and interpreter of hieroglyphics,
a remarkably agreeable man he is, and his conversation shows great ability. It is superior, I think, to
his book. My first introduction to him was on the
2nd.

Went again by appointment, at one o'clock, to Humboldt's, to meet M. Schacht, a young man who has made important researches in physiological botany. He brought a large number of his drawings of microscopical details of structure. Humboldt estimates him very highly as an observer, and indeed spoke of him to me as the first botanist now living in Germany. His drawings are certainly most beautiful, especially the anatomical details of the Coniferæ, Rafflesiaæ and Balanophoreæ. He seems very moderate and intelligent. Humboldt's activity of mind is truly astonishing ; no department of science seems to escape his attention, and in every branch, if he has not himself made original researches, he seems acquainted at least with all that is going on, and qualified in some degree to estimate the merits of the workers. He seemed to take much interest in Schleiden's theory of the action of the pollen, the truth of which theory he considered to be fully proved by M. Schacht's researches. The theory seems to amount to this— that the germ of the future plant is really contained in the pollen-tube, and that the " sac embryonaire " of the ovule merely furnishes a receptacle for it.

June 6th.

Went to M. Schacht's lodgings to see some of his preparations, exhibiting the pollen-tubes under the microscope. They are very curious and seem to establish the accuracy of his drawings. As to the theory which he thinks they establish, it is too

1855. difficult a subject, and one which I have hitherto too little studied, for me to pronounce an opinion on it.

In the evening went to a party at Dr. Lepsius's, where I had some talk with Dr. Peters, the African traveller, a very pleasant man. He is engaged in preparing for publication the zoological results of his travels, and has published one volume containing the mammalia. He told me that he spent six years in his travels, and has now already been engaged for seven years in arranging and publishing his collections.

The giraffe, he told me, does not inhabit the part of Africa which he explored, but he does not doubt that the giraffe of Nubia is the same species with that of Southern Africa. Some species of antelope are common to Equatorial and Southern Africa, as well as the lion and several other carnivora.

The hippopotamus is very common in the rivers of the Mozambique coast, and he and his people killed many; they fight fiercely, and dangerously when they cannot escape by diving. He seems to think it probable, though not quite certain, that the hippopotamus of the Nile is the same species with that at the Cape, but, he says, there is a much smaller kind found in Western Africa, which is certainly a good species.

June 7th.

A farewell visit from M. de Humboldt; extremely pleasant. He had already very kindly made me a present of his " Mélanges de Geologie," with the Atlas of Views in the Andes. He talked very

agreeably of various eminent men whom he had 1855.
known, of Warren Hastings, and especially Canning,
with whom he was intimate, and whom he described
as having a peculiar charm in his manners and
conversation. Canning, he said, was not in the
least Frenchified, as accomplished and agreeable
men of some countries, Russians in particular, are
apt to be ; he retained all the characteristics of an
Englishman, and at the same time was as agreeable
as a man can possibly be.

When Canning was appointed Governor-General of
India, one of the first things he did during his short
tenure of that office, was to write to Humboldt, asking
him to accompany him to India. And this was at a
time when the East India Company were particularly
jealous of admitting any foreigners to their possessions ; and they had even specially protested against
allowing Humboldt himself to travel there ; fearing
no doubt that he might *show them up* as he had
shown up the Spanish Government in his '' Essai
Politique sur la Nouvelle Espagne.'' Humboldt
however spoke with great candour, perhaps with too
much indulgence, of the East India Company's
government, though he characterized it very justly
as a *gouvernement proconsulaire.*

He told us he had been present at the trial of
Warren Hastings, and had heard Sheridan, Fox and
Burke ; and he had been present also at the trial
of Queen Caroline.

LETTERS.

Hotel des Princes,
Berlin,
June 3rd. 1855.

My Dear Father,

1855.   I hope the fine weather you mention in your letter has continued with you, as it has with us here, with the exception of one or two cloudy or showery days, and one heavy thunderstorm, we have had beautiful weather ever since the 21st of May: and especially since we returned to Berlin it has been very hot. We have plunged at once into summer, the trees are in full deep foliage, and the Lilacs and Horse Chesnuts in gorgeous blossom. —The 21st, the day before we started from Berlin, we saw a grand review of the whole garrison, twelve battalions of infantry, thirteen squadrons of cavalry, and fourteen batteries of eight guns each. There were no manœuvres, the several regiments with their bands playing merely marched in succession past the king, who, with his brother the Prince of Prussia, and several general officers, was on horseback in the square of the Opera House, in front of the statues of Blucher, Gneisenau and Yorck. The day was beautiful, and the spectacle a very pretty one, and we were well placed for seeing it. The soldiers, be it observed, were not veterans, but for the most part young men going through their required three years of service: nevertheless, to my unpractised eyes, they made a very good appearance. I wish I could have known what you would have thought of them.

On the 22nd we went to Breslau, and remained 1855.
there three days, for the sake of conversing with
Professor Göppert, who is famous for his knowledge
of fossil plants, and of seeing his collections.

Breslau, though it has no regular *sights*, is worth
seeing on its own account: it is the most curious,
quaint, picturesque old town I have ever seen,
surpassing even Brunswick in its lofty fantastic old
houses and savouring thoroughly of the middle ages.
The public walks which entirely surround it,
occupying the site of the old fortifications, are
extremely pretty and agreeable. It is said, however,
to be an unhealthy town. The country about it,
and between it and Dresden, has been the scene of
as many battles, I suppose, as almost any tract in
Europe. We spent three days and-a-half at
Dresden very pleasantly, and passed a good part
of each morning in the gallery, which is indeed a
glorious collection. We were just in time, as, the
very day after we came away, the gallery was to be
closed, and the removal of the pictures to the new
building to commence. Neither should we have
been able to see them on Whit Monday, but for the
special favour of the director, Professor Schnorr,
to whom we had an introduction, and who, on his
visit to England some years ago, had moreover
known Mr. Horner. He was most obliging and
kind to us.—The Madonna di San Sisto is a truly
divine picture. I have no hesitation in saying that
(not speaking for the present of frescoes) it is by far
the finest, the most glorious picture I have ever
seen. There is something indescribably solemn and

1855. awe-inspiring in those deep, earnest, *far-seeing* eyes
of the Mother and Son, of which no copy, nor print
that I ever saw, gives at all an adequate idea.
Both seem to look far into the depths of futurity,
but the Mother with a more mournful expression,
the Son with a more grand and majestic resolution.
It is a picture that produces upon one somewhat
of the same impression that a noble Gothic
Cathedral does. It is a pity that it is not placed
in a room by itself. It is very thinly painted, and
the colouring faint, so that it ought not to be in
company with more gaudy works. There is a new
engraving of it by Steinla, which gives in some
respects, I think, even a better idea of it than
Müller's famous one, and it is less expensive.
Holbein's Madonna far surpassed my expectations:
it is not sublime like the San Sisto, but most
beautiful, a far higher degree of beauty in the Virgin
herself than I had at all supposed any artist of that
school to be capable of: and most admirable
colouring. The Correggios, on the other hand, disappointed me more than anything else in the Gallery,
—with the exception, however, of The Magdalen,
which is lovely. The collection altogether is an
immense one,—nearly 2000 pictures, and many of
them no doubt very trashy, but it is very rich and
complete in nearly all schools.

Dresden is a very fine and very pleasant town, but
not so readily characterized in a few words as either
Breslau or Berlin, for it is neither a thoroughly
quaint and picturesque old town, nor a regular and
symmetrical new one. I should think it must be a

pleasant residence. It has a great advantage over 1855.
Berlin in its situation, the country about it being
very cheerful and pretty, and the river beautiful.
We had not time to go to the Saxon Switzerland,
but took a pleasant afternoon drive to Tharandt,
about ten miles off, through an extremely pretty
valley. I found also at Dresden a very interesting
collection of fossil plants in the museum, under the
charge of a very able and well-informed professor.

We returned hither on the 30th of May, and are
spending our time as before, both pleasantly and
profitably. There could not be a better position for
study. I have liberty to consult books in the library,
and have thus had the opportunity of consulting
some rare and costly ones, which though well known
to me by name, I had never before met with.

We have made a new acquaintance, that of
Dr. Lepsius, the great Egyptian scholar, a remarkably agreeable man. I have seen a private collection
of modern pictures, chiefly German and Flemish,—
Mr. Wagener's,—in which are some very interesting
things : in particular, that picture of " The Warrior
and his Child," by Hildebrandt, of which you have
the print. Very beautiful it is, and excellent in
colouring as well as in other respects. "A Robber,"
by the same artist, is very fine. There is also the
" Don Quixote," by Schroeder, of which you have a
print; and several other excellent things, among
them some beautiful landscapes by Koekkoek, a
modern Dutch painter. Altogether the collection
gives one a high idea of the state of art in these
countries.

1855.   *(June 4th).*   Yesterday was one of the most oppressive, suffocating days I ever felt, but towards evening there was a fall of rain, and this morning is fresh and bright and pleasant.

At this distance it is of little use to speculate on the course of events in England.   I was surprised at the greatness of the majority against Disraeli's motion, but I suspect it rather implies dislike and distrust of Disraeli, than real confidence in the present Government.   Gladstone's speech was a remarkable and significant one, and makes me very glad that he and his party are out of office; long may they continue so!   Even Bright and Milner Gibson would hardly, I think, be worse advisers on the great question of war or peace.

The news from the Crimea begins to look more favourable.   Pelissier has begun with something like spirit, and really does seem inclined to do something; and the successes at Kertch and in the Sea of Azof are not only good in themselves, but really do look like the beginning of a more active system. We may hope now to hear less of conferences and more of successes,

Pray give my love to Emily (with many thanks for her letter), and to Henry and Cecilia if they are with you.

Believe me ever your affectionate Son,
C. J. F. BUNBURY.

Barton,
July 23rd, 1855.

My Dear Charles,

Thanks for your letter of the 20th.

Tom Scott is so busy superintending the hay-making, after the soaking my hay has had that I cannot get hold of him to-day to settle about a cow: but you shall have one very shortly.

The conduct of Austria has certainly been shabby, but shabbiness has long been one of the characteristics of the Austrian Government. It seems to be rather the result of timidity than of treacherous design ; and history shows that not only her statesmen, but likewise her generals, were vacillating and timid, except when her armies were led by the Savoyard Prince Eugene, or the Scotchman Loudan.

We expect poor Cissy by the express train this evening. Henry embarked on Saturday ; but whether the Orinoco has actually sailed I do not yet know. Henry is doing his duty honourably, and we must hope the best ; but Sebastopol is a fearful place.

Much love to Fanny, and thanks for the Dresden catalogue.

Most affectionately yours,
H. E. B.

Mildenhall,
July 29th, 1855.

My Dear Mr. Horner,

I thank you very much for your Memoir on Egypt, which I received from Mrs. Horner. I

3 D D

1855. have read it through with attention, and think it very clear, interesting, and instructive. I am particularly pleased with your preliminary view of the physical and geological structure of Egypt in general, which appears to me a very masterly piece of physical geography. Your researches have been so well planned and carried out on so noble a scale, with so much perseverance and such elaborate care, that I am extremely glad they are to be given to the scientific world, through a medium which will attract attention to them, and do them justice. I shall be eager to see the subsequent memoirs in which you will develope the theoretical conclusions that you deduce from the facts hitherto ascertained.

I am much obliged to you also for letting me see the examination papers of the Edinburgh Academy. I am not very conversant with modern Latin verses, but the exercises in that and other kinds of composition appear very creditable, and the passages selected for translation indicate a high standard of scholarship. I congratulate you on having found that institution in such a flourishing state ; it must be a great satisfaction to you to see its continued success and usefulness. I am very glad you are passing your time so pleasantly at Edinburgh, as you appear to be by what Mrs. Horner tells us. Although no doubt you miss many old friends there, I can well understand how much you must find there to please and interest you, and how much pleasure you must have in renewing old associations and re-visiting well-remembered scenes. We meanwhile, are thoroughly enjoying the society of 1855. Fanny's dear mother and sisters ; and it is indeed a great pleasure to receive them under our roof, and they are as agreeable as they always are. My darling is tolerably well ; but not as strong as I could wish ; perhaps the moist relaxing weather is against her.

I am exceedingly pleased with Hooker's " Flora Indica," which I am studying carefully ; it is a most masterly work, and will, I think, not only support but even heighten his previously great reputation. I am also going carefully through Geinitz's new work on the " Fossils of the Coal Formation in Saxony," and I find it so important for the knowledge of fossil plants, that I intend to send a somewhat full account of it to the Geological Society's Journal. His views appear to me remarkably sound, and I can answer for the correctness of his plates and descriptions, having seen the original specimens in the museum at Dresden.

Ever your affectionate Son-in-law,

C. J. F. BUNBURY.

Mildenhall,

September 7th, 1855.

My Dear Katharine.

I have a pleasant letter to thank you for, which I had intended to answer sooner, but I have latterly had very little time to spare. For the first fortnight or so after we returned home we were very quiet, and I worked steadily, especially at reading and making copious notes from a new German work, by Professor

1855. Geinitz, on the "Fossil plants of the Coal Formation
of Saxony;" this I bought at Dresden, and found to
contain such valuable information, that I have been
going regularly and carefully through it.   Since
then, and especially of late, I have had many dis-
tractions and interruptions : visits to Bury, to
Barton, to Hardwick, to Ely, &c., &c.   When
Sedgwick was here, and again when Mr. Gibson* was
here, Fanny was anxious to shew them everything
that was to be seen in the country, far and near, but
she was not able to go out much herself, and
Susan and I did the honours of all the neighbourhood
for them.   Indeed during the four days of Gibson's
visit, I had not a minute to myself, and he too was
kept on nearly as active service as the Queen during
her Parisian visit !   We took him to Barton one
day, to Ely the next, to Ickworth the third, and to
Cambridge the fourth.   He was exceedingly agreeable, and seemed to enjoy racketting and sight-
seeing ; but I mention these things just to show
you how my time and attention have been
dissipated, and that I have not had so much leisure
for botany as you might suppose.   We have now
the pleasure of having the Pertzes with us, and
very pleasant they are, and I am delighted to see
Leonora looking so well and so strong after her long
journey.

Now for botany.—Much of my time has been
occupied in arranging and examining two large
collections which I received in July :—a large parcel
of Indian plants from Joseph Hooker, and a most

Sir John Gibson, the celebrated sculptor.

beautiful set of Ferns from various countries, 1855. duplicates from Sir W. Hooker's herbarium. These would delight your eyes. There are among them 25 species of Adiantum, as many of Davallia, nearly as many of Trichomanes, and so on of many other genera; many wonderfully beautiful, and some very rare. I have arranged them all so far as placing them under their proper genera, but have not yet had time to give many of them a thorough examination. Hooker's Indian plants are mostly from Sikkim, and the Khasia mountains; not so well dried as some I have seen, but many of them very interesting; among the number are 16 species of Rhododendron, nearly as many of Vaccinium, 9 or 10 Oaks, some very fine Magnolias, a most curious Nepenthes, a great number of Potentillas and other Rosaceæ, and numerous species of Cherry. I suppose you have seen his "Flora Indica," a most admirable book; the Introduction to it is one of the most interesting and valuable botanical treasures I have ever read. I make it my daily study and text-book. The only think to be regretted about it, is that, even supposing him and Thomson to live to the age of Humboldt, and to work assiduously all their days, they can hardly by any possibility complete the work.

I have had another job in hand, which I have just finished after a fashion: no less than the re-arranging of my whole collection of plants,—the cryptogams excepted. As the collections increased I found the geographical arrangement involved an excessive waste of space, and also of time and

1855. trouble whenever I wanted to compare the nearly allied species of different quarters of the world, or (as sometimes happened) specimens of the same plant from distant countries   I therefore determined to throw the whole together into one general systematic arrangement, as I had already done with the Ferns ; and this arrangement I have just completed in a rough way : but it has cost me a good deal of time.   I have thus not been able yet to begin upon the Cape plants which you kindly gave me.

We have almost decided to go to Malvern as soon as the Pertzes have left us.   A change to a better air for a time is strongly recommended on Fanny's account, and there is hardly any better air than that of Malvern ; it is a place too that I have a great wish to see again, and where I should find a great deal to interest me, particularly in Geology ; it would suit me much better than any other place that is recommended for air, and I hope two or three weeks there will quite set Fanny up, and enable her to face the winter at Mildenhall.

I hope you have enjoyed your visit to Scotland. I am very glad to hear that your boys are so flourishing, and glad too that Leonard has not forgotten Uncle Bunbury.   Give my love to both of them and to Harry.

Ever your affectionate Brother,
C. J. F. BUNBURY.

Bath Cottage, Malvern Wells.

October 7th, 1855.

My Dear Lyell,

We made out our visit to Mr. Symonds, on Friday, very satisfactorily, the day turning out much finer than it promised at first. I had a capital geological walk with him : he took me first to the Keuper quarry near his house, where he showed me the bone-bed, the nodules of argillaceous limestone with Posidonomya, and the fragments of coaly matter in the sandstones ; in these we saw also some larger pieces of vegetable matter, wood and bark, perhaps coniferous, but quite undeterminable. Next he took me to the house of a lady (Miss Edwin) who has a fine specimen of a Fern (or Cycad, according to Brongniart), Otopteris obtusa, from the Lias of this county. I dare say you saw it, or at least heard of it. It is a good specimen, but I have a better from Baireuth, from a formation which some of the German geologists refer to the Keuper, and some to the Lias. Then we went on to the pass between the Keysend and Raggedstone hills, where he showed me the very remarkable section in which the contact of the *Hollybush* sandstone with the trap rock, and the metamorphic effect produced on the former, are so well displayed ; and we ended with the black *Olenus* shale, in which the trilobites are *not* now to be found. I have seldom had a more instructive day of field geology ; nor could I have had a better guide to it. I wrote you word on Wednesday that an intended expedition to

1855. Eastnoor with Mr. Symonds, had been baffled by the rain.

However, with great spirit and good nature he came the next morning, theweather promising a little better; and though this promise proved deceitful, and the day turned out very bad, we yet saw something. He explained to me the relative positions of the Ludlow, Wenlock, and Caradoc rocks, their geographical positions, I mean; and we collected some good specimens of characteristic fossils. Certainly this is a most interesting country; it is marvellous what a quantity of good geology is brought together within the space of a few miles. I have not however, seen anything of the least importance in the way of fossil plants, with the exception of the aforesaid Otopteris. The bed of the Keuper from which Mr. Symonds, obtained the specimen of Calamites (?) that he sent to London is now he tells me, buried by rubbish.

I take very much to Mr. Symonds; he is as pleasant a companion in a walk as I have often met with, and his activity of mind, candour and freedom from prejudice, and ardour for knowledge, are very remarkable. I wish we had anybody like him in our part of the country !

Much love to Mary, and to all the house of Horner.

Ever affectionately yours,

C. J. F. BUNBURY.

Bath Cottage, Malvern Wells,   1855.
October 14th, 1855.

My Dear Mr. Horner,

This is certainly a most lovely country; I have hardly ever known anything more enjoyably beautiful than these hills with the scenery they command, and we do enjoy them to the utmost. Hardly a day passes that we do not reach the top and feast upon the views, so variously beautiful on the two sides and so exquisitely diversified by the accidents of light and shade, and the different states of the air on different days and at different times of the day. The delightful bracing air adds to the enjoyment, and I think it would take one long to get tired of such a country.

We revel in geology, and Fanny is taking quite a keen interest in it, and from her assiduous study of Mr. Symonds' little book, and from his conversation, is *getting up* the whole Silurian system capitally.

To be sure it is a most fascinating country for geology, a perfect compendium of the older fosiliferous formations, as instructive a display of them as the Isle of Wight is of the younger rocks. We are fortunate too in having such a well informed and pleasant guide and companion as Mr. Symonds. You will probably have seen Fanny's letter to Katharine in which she gave a capital account of our pleasant expedition with him to the obelisk hill in Eastnor Park. We collected that day some good fossils of the lower Caradoc, and saw in the Gullet-wood Pass, an interesting example of apparently metamorphic structure; a rock which one cannot

1855. but call true well characterized mica-schist, yet which certainly appears to be the old *Hollybush* sandstone, altered by felspathic dykes.   You have probably seen the remarkable example a little further south on the south side of the Ragged-stone hill, where this conversion of the *Hollybush* sandstone into micaceous schist by an intruding mass of greenstone is so clearly traceable.   But the case I mentioned before is considered by Mr. Symonds as peculiar and exceptional because he has nowhere else seen any metamorphic action distinctly traceable to the true felspathic Malvern syenite.

No one, I think, can examine these hills geologically and fail to be struck with the excellence of your description of them written so many years ago.   I find but one opinion as to the accuracy of that memoir; and whether as to the physical geography of the hills, or the careful and exact mineralogical description of the rocks, or the philosophical spirit of the whole, it is a masterly work. You know I am not given to paying compliments, and I say this only because I am strongly impressed with the truth of it.   Indeed the geology of this region has been most admirably worked out ; few districts in England better I should think.

Yesterday, a beautiful day, I took a long walk by myself (Fanny not being disposed to go so far) to Eastnor Park, examined two or three quarries in the Wenlock limestone, and collected a few fossils ; but I have not yet been able to find a trilobite. Tell Joanna that I hope to be able to give her a few Silurian shells.

Fanny, I am happy to say is very well and 1855.
visibly benefited by the Malvern air, and it delights
me to see how thoroughly she enjoys the place.
We continue to like our lodgings, diminutive
though they be. Very sorry I shall be when the
time comes for leaving this charming country;
but I hope we shall have gained enough to
preserve us from complete Suffolkation during the
winter.

I have been very glad to hear such comfortable
accounts of you all, and in particular that Leonora
is so well. We hope soon to have still further
good accounts of her.

Pray give my love to Mrs. Horner, to all the
sisterhood, and to Charles Lyell, Harry and Pertz.

Ever your affectionate Son-in-law,

C. J. F. BUNBURY.

———

Abergwynant,
November 7th, 1855.

My Dear Katharine,

I have long owed you a letter, and now we
are settled here for some days, and a wet and
stormy day affords a good opportunity for letter
writing, established as I am in a comfortable warm
room. I was much obliged to you for your letter,
and very glad to hear that you have got your Ferns
back from Kew, named on such good authority, and
that you are setting to work to arrange them. I
shall like very much to go over them with you, and
shall be very much obliged to you for the names of
the Indian ones, most of which I fancy are not

1855. described in any work I have access to. I shall
have much to do with my Ferns during the winter.
I have seldom passed a pleasanter month than that
which we spent at Malvern; it is a charming place,
really one of the most enjoyable that I know, so
beautiful and so accessible; and I do not know a
more delightful air. We both left it with very great
regret. I had filled a box with rock specimens and
Silurian fossils.

In botany my collections were much less considerable; the season was too late for the generality of
flowering plants, though the Furze was in a blaze
of blossom all over the hill sides, the Fox-glove still
showing here and there a spike of flowers, and
the pretty little Corydalis (or Fumaria) claviculata
in full perfection, climbing among the Furze bushes
with its delicate, finely cut pale glaucous green
leaves, and neat little white flowers. The limestone
slopes and woods on the west side of the hills are
richest in curious plants, but nothing remained
there except the Chlora, still in flower in some
places; Astragalus glycyphyllos in seed, and
Hypericum Androsaemum with ripe fruit. In Mosses
the Malvern hills appear to me remarkably poor,
whether it be owing to the narrowness of the range,
and its insulated position, which render it uncommonly dry, or whether it be the want of extensive
masses of rock and of deep ravines, or whether the
nature of the rocks themselves is unfavourable;
certain it is that there seemed to me to be marvellously little variety of Mosses, either on the
hills themselves, or in the woods at their bases,

and many even of the common Mosses of hill 1855.
countries appeared to be wanting. Ferns too are
comparatively few, and the only one at all out of the
way, that I met with, was Aspidium Oreopteris,
which I dare say you know well in Scotland. *Here*,
I expect to get plenty of Mosses, as I know from
former experience that it is a good place for them,
and the wet weather is favourable for a Moss
harvest. I will get what I can for you, as I know
your collection of Mosses was partly spoiled during
your absence in India.

It was a great addition to the pleasure of our stay
at Malvern, having frequently the company and
guidance of such an agreeable man and so good a
geologist as Mr. Symonds. He is really one in a
thousand; with such an eagerness for knowledge,
such an enlightened and candid mind, and a
contagious ardour and activity like Charles Lyell's,
that it does one good to be in company with him.

I believe we were mutually pleased with each
other, and I learned much from him. I was able in
return to give him some information that was useful
to him. Fanny was as much pleased with him as I
was. We left Malvern (dear Malvern!) on the 25th
of October, Mr Symonds joined us near Ledbury,
and we went together to Hereford; we had intended
to visit the Woolhope Valley on our way, but the
weather proved too bad. That same day, as I hear
from my Father, there was a perfect deluge in
Wales, which made most of the streams overflow,
carried away bridges, and did much damage. The
next day proving fine, Mr. Symonds and I went to

1855.   Dormington, and explored part of the Woolhope
valley, and the next day we went thither again, a
party of four, including Fanny and the Dean.

Examined the Dormington quarries, and collected
many fossils. I enjoyed those two excursions very
much.

The Valley of Woolhope *(a valley of elevation,*
geologically speaking), is strikingly beautiful and
picturesque,, and of a *very peculiar style of beauty,*
and its geology is remarkably interesting. We
remained at Hereford the following Sunday and
Monday, miserably cold days : attended service in
the Cathedral, where I was half-killed with cold ;
and saw my old tutor, Mr. Mathews. The Dean
and Mrs. Dawes were very kind and pleasant.

We spent five days at Bath, pleasantly enough.
Hanmer and his wife were very kind and affectionate, and very glad to see us, and so were the dear
children, whom we are very fond of. I see Fanny
has written you some account of our proceedings
there. But the weather was wretchedly cold, and I
felt stupid and lazy, as I am apt to be in cold
weather. On Monday we had a long fatiguing
journey from Bath to Shrewsbury, and yesterday we
arrived here long after dark, very tired, in the midst
of a most furious storm of wind and rain. To-day
the weather is little better, so that the place does
not appear to advantage. My Father seems pretty
well, and Lady Bunbury much as usual. Cecilia is
in a separate house at some little distance, and
quite laid up, poor thing, so I have not yet seen
her.

I have been delighted to hear such excellent 1855.
accounts of Leonora and of "Miss Pertz," who
I hope will continue to thrive as she has began.

I trust your boys are flourishing.

Pray give my love to your husband, to Charles
and Mary, and to the house of Horner, and believe
me.

Your very affectionate brother,

C. J. F. Bunbury

Mildenhall,
February 3rd, 1856.

My Dear Katharine,

Many thanks to Leonard for his message; 1856.
give him my love and tell him and Frank not to
forget Aunt Fanny and Uncle Bunbury, and to
come here again in summer, when they may gather
wild flowers.

You will have heard from Charles and Mary all
about their visit to Barton and Ickworth, and the
lecture, which was really a splendid one; almost
the best I think that I ever heard him give;
excellent in every respect. It was wonderful what
a mass of knowledge he brought into that space,
and all so well arranged, all bearing full upon the
point, and forming a close, compact chain of
reasoning. It was really a masterpiece. And
though, in that crowded room, there were perhaps
comparatively few who could thoroughly follow the
whole of it, I think its merit was generally felt. I
need not say that Charles and Mary's visit to us,
short as it was, was a great treat to us, and that we

1856. enjoyed it thoroughly; and I think they were pleased with Barton. From the abundance and beauty of the evergreens, those grounds look well at all seasons; and the conservatory was very gay indeed, with the various colours of the Camellias, the profuse, bright, yellow blossoms of the Jasminum nudiflorum, and the rich purple of Cinerarias and the Rhododendron Daüricum.

I spoke to the gardener there about some slips or cuttings of Ferns for you, as they have a much greater abundance than we have; but he thought they could not travel safely in such cold weather.

To-morrow I shall be forty-seven years old! A long space of time to look back upon; and, on the whole as happy a life, I take it, as most men have enjoyed; and if I had the choice of living it over again, there are but few things *external* to *myself* that I could wish altered. When I compare my opportunities with what I have done, I certainly have no room for pride or vanity; but I may hope that in spite of Dean Barnard's doctrine, I am not yet too old to improve or to learn.

Now that I have got my lecture off my hands, I feel more at liberty, and shall return with zest to the study both of Ferns and of Cape Plants, as well as to my general work on fossil plants and the examination of the fossil leaves from Madeira. I am in no fear of wanting occupation. I am still going on with Macaulay, which interests me extremely, but it will not last me much longer, and when I have finished those volumes, I mean to read the same period in Burnet.

We have accepted an invitation for the early part 1856. of next week, to Mr. Birch's, at Wretham, beyond Thetford; after we return I hope to be quiet for some time. There are several things I wish I had shown you when you were here, but I was *bothered* by the preparation of my lecture, and some other drawings that I had to get done before a given day. I hate being tied down to time. You must come again.

My love to Harry and your fine little fellows, and to all the party at Queen's Road West. I hope to see Joanna and Charles Pertz here.

Ever your very affectionate Brother,

C. J. F. Bunbury.

Mildenhall,
February 5th, 1856.

My Dear Mary,

Very many thanks for your kind letter, and your present of Rogers's portrait, which is exceedingly like. I do not know whether, by choosing *that* as a birthday present, you intended to convey a wish or expectation that I may live to be as old ; but at any rate I thank you. I was much entertained by the anecdotes of Leonard and Frank. I am much disposed to agree with you as to Glencoe ; though I cannot quite consider King William "as an accessory before the fact," but you will find when you read further on, that he certainly became an " accessory after the fact," by his criminal indulgence to the Master of Stair (the real murderer) after the business had been thoroughly investigated by a Commission,

3 E E

1856. and the facts ascertained beyond a doubt or cavil. Even to the Master of Stair, Macaulay is perhaps too lenient in his explanation of his motives. I apprehend the fact was, that the Master looked upon the Highlanders very much in the same light in which some Englishmen, even in our day, view the Caffers. But how admirably well Macaulay tells the story; it would be difficult for any writer to excite stronger feelings of abhorrence against the perpetrators than he does. I am now within a hundred pages of the end of the volume, and I only wish there were two more volumes to read. Scarcely anything is better told than the plot for the assassination of William in 1696; it is as interesting as any novel; it might furnish a useful lesson to *desperate* politicians, to observe how the king's popularity was revived and his government strengthened by that unsuccessful plot.

I hope we shall not drift into a war with America, it would be a great misfortune, and a most useless and causeless waste of blood and treasure. It would be like an unreasonable duel between two kinsmen, in which both may suffer, and neither can gain anything. Unfortunately there seems to be a good deal of irritation on both sides; on our side mainly owing, I think, to the loudly expressed Russian sympathies of a part (at least) of the Americans. I can hardly suppose that the great body of the American people wish for such a war; but the president and his faction have been as mischievous as *The Times*. If the two nations are drawn into a war, President Pierce and the editor of

The Times will have to share between them the 1856.
guilt of blood.

General Simpson, who called here yesterday, is
inclined to think that the Russian professions of
desire for peace are a mere feint ; and he says that
an armistice would be advantageous only to the
Russians.

I have not *yet* done with my lecture, for at the
petition of the Mildenhall Institute, I have consented
to repeat it here ; day not yet fixed.

With much love to Lyell.

I am ever your very affectionate Brother,

C. J. F. BUNBURY.

Mildenhall,
February 6th, 1856.

My Dear Father,

I have been more gratified than I can
express by your truly kind and affectionate
expressions towards me in the note which I have
received from you this morning. Most heartily do I
thank you for them. Your affection and approbation
are indeed very dear and precious to me, and I hope
and trust, with God's help, that I may always
continue to deserve them. I have many and many
blessings to be thankful for, and very high among
them I rank the love and kindness of so good a
Father. If there is any good in me, I feel it is very
much owing (under God) to the example of yourself
and my dear Mother, to your care of my education,
and to the advice I have at all times received from
you. It is a great comfort that you are comparatively so free from pain, and I most sincerely trust

3 E E 2

1856. that you may yet live many years, with the same
exemption from suffering, and with your faculties
equally well preserved.

Pray give my best love to dear Emily, with my
hearty thanks for her very kind note ; I will write to
her very soon.

Ever your truly affectionate Son,

C. J. F. BUNBURY.

P.S.—General Simpson, who visited us the day
before yesterday, thinks, like you, of the prospect of
peace, and doubts the sincerity of the Russians;
he seems to think that one of their objects is to
secure an armistice, which would be advantageous
only to them.   All accounts seem to agree as to the
extraordinary eagerness of the French (people as
well as Government) for peace ; which is something
quite new for *them.*   I have some hope, however,
that our Government will not allow themselves to be
hurried on this account into a hasty or inglorious
peace ; and if they are firm, I am confident they
will have the support of the nation.   On re-consider-
ation, I like the Queen's speech much better than I
did at first ; at least, that part of it relating to war
and peace ; I think its tone firm and dignified.

Henry appears, from what Cissy writes, to
anticipate a campaign in Asia Minor this next
summer.   I do not apprehend that there would be
anything particular to fear from the climate, in such
a case ; the coasts of Asia Minor indeed are very
malarious, but I fancy the Russians have no hold on
the coast anywhere ; and the whole interior of the

country, I conceive, is a high table land, with a 1856. general elevation of some thousands of feet above the sea-level.

I am drawing near to the end of Macaulay's fourth volume, and only wish there were two more volumes to read. He has a marvellous power of interesting narrative ; he makes even the financial difficulties of 1696 interesting, by his way of treating them. Once more, I am ever your very affectionate Son.

Mildenhall,
February 19th, 1856.

My Dear Emily,
I thank you for your pleasant letter of the 13th, and am glad to find you are so busy with Mosses. They are indeed fascinating little creatures and you have them in their glory at Abergwynant ; I have hardly anywhere seen them so beautiful. I wish you could find Hookeria lucens (ci-devant *Hypnum lucens)* which ought to grow in the same sort of places with the Bryum punctatum, but which I have not been able to find at Abergwynant. Of the two Mosses you sent me, neither is the strictly normal or typical state of Dicranum scoparium ; the larger approaches to the form which many botanists call Dicranum majus ; the smaller is a variety not uncommon, but which has not (as far as I know) received a name, though it is rather peculiar in the interrupted or tufted growth of its leaves.

Our visit at Wretham turned out very pleasant ; Fanny has written Cissy a capital account of it,

1856. and of the people we met there, which I hope you have seen. I like the ladies of the family very much, and Mr. Birch is a very pleasant lively old gentleman, wonderfully active and full of spirit, and much more to my taste than those very practical men often are. He has, as you say, a fine head. I was rather interested in the peculiarities of the mere he has drained ; the occurrence of a distinct bed of undecayed and well-preserved Moss beneath fifteen feet of black, peaty mud, showing no trace of Moss, but containing numerous red deer's horns, is curious.

I read Burnet nearly at the same time of life as you did, namely at twenty-three, but have never looked at him since, so I thought it time to rub up my recollections. His slipshod style does not read well after Macaulay, but his simplicity is amusing, and one feels great confidence in his honesty, though not quite so much in his judgment. It is very odd that in giving an account of the siege of Derry, he should make no mention whatever of Walker. As for poor " Haydn's Life," I have seldom been more inter-ested by any book ; it is such a perfect picture of a human mind, such a thorough and undisguised laying open of his character. For the frank dis-closure of character, motives and feelings, and weakness, I do not remember anything like it, except Pepys ; but it excites very different emotions. Haydn's was certainly a mind of un-common power ; with what vigour he writes, and with what masterly touches he brings out the

characters of those he is brought in contact with ! 1856.
He seems to have been unsuccessful as a portrait
painter with the brush ; but his portraits with the
pen appear to me quite masterly.   But it is a
very melancholy book ;   his faults and weaknesses,
poor fellow, were many and obvious enough, but
his sufferings were surely much more than in
proportion.   What is to me almost the saddest
part of all is, that with all his powers of mind, all
his energy and perseverance, he did not succeed
in making a great name, nor as it seems, in
becoming a really great painter.   It would seem
as if he must have mistaken his vocation.

Mildenhall,
March 28th, 1856

My Dear Katharine,

Many thanks for your letter.   I am always
very glad to hear from you, and to have news of
your botanical proceedings, and most happy to help
you whenever I can.   I hope you will have already
received a copy of my Madeira paper.

Raddi's name is familiar to me ; he was an excellent botanist, who spent some time in Brazil
at the expense of the Grand Duke, and published
an important work on Brazilian Ferns (which I have
not yet been able to get) ; he afterwards went on
a scientific mission to Egypt, and died there.
There is a monument to him in Santa Croce at

1856. Florence.   It will be very interesting to have
specimens collected and named by him.

The mildness of the first two months of this year
was favourable to Mosses, and I have found Hypnum splendens, squarrosum, and Schreberi, in
good fruit for the first time here ; but splendens is
not yet ripe.   They have their own times and
seasons for fruiting, though this is generally too
much neglected in the books : thus, in Wales in
November, Hypnum proliferum had its fruit nearly
ripe, while splendens growing with it, and in equal
vigour, was only showing its stalks and veils, the
capsules not being even formed.   If either splendens
or Schreberi should happen to be among those of
your specimens which have been damaged, I will
keep some for you.

I have been working again at Ferns, and have
gone regularly through the Polypodeæ, but I do not
yet see much daylight as to the principles of distinction of genera.

I am very sorry dear little Arthur is still so
delicate.   The season is a very trying one, even to
strong grown-up people, and very hard for those who
are at all delicate, at whatever age, to stand their
ground at all.   I caught cold at the assizes on
Tuesday,—a most bitter day it was at Bury,—but it
is not one of my very bad colds.   We came back
yesterday, as my Father had written word that they
might be at Mildenhall to-day, but now we find they
have put off their departure again for a few days ;
and no wonder, considering the weather.   A horribly
cold month it has been, indeed I really felt it

colder the other day at Barton than when we were 1856.
there in January. Yet after all, this season is earlier
than the last ; the Ribes sanguineum, for instance,
at Barton was in flower by the beginning of this
week, whereas last year it was hardly even showing
a bud at that time. Our rooks, however, have
been uncommonly late in building this year, so that
we began to be afraid they had deserted us,—but
they have built at last.

It so happens this is the first *March* we have
spent at home since we have been married, and we
amuse ourselves with registering the first appearance
of flowers and leaves and birds and insects. I long
for the appearance of really mild Spring weather,
which would be the best of medicines for Fanny.
She is contumacious against doctors, and sets the
whole brood of Asclepios at defiance.

With much love to your husband and your dear
little boys, and to all the party at 53.

Ever your very affectionate brother,

C. J. F. Bunbury.

Mildenhall,
April 7th, 1856.

My Dear Katharine,

I have put up a few Mosses for you, of
which I inclose a list ; and they are ready to be
sent whenever an opportunity occurs, but the parcel
is rather too large to go by post. There are
twenty of the species included in your list, and a
few exotic varieties which I thought you might be

1856. glad of. Your letter set me to work again at my
own Mosses, which I had long neglected: and I
have been going on with my arrangement and
catalogue of them, and studying Bridel's Bryologia,
which is an excellent text-book, full of information,
though I often think him mistaken in his views,
especially as to species. I believe Nature meant
me for a curator of a museum, or some such thing,
I have such a delight in arranging and cataloguing;
indeed I am too fond of it, for it often leads me
away from more philosophical occupations. However, as I never implicitly follow any published
arrangements, nor put down any species in my list
without having examined and compared it, this
occupation at least makes me more thoroughly
acquainted with the objects. To return to the
Mosses, with regard to species, the proper allowance
will probably lie between the number admitted by
Bridel, which is doubtless too high, and that of
Hooker and Taylor, who certainly do seem to me
to have in many instances united things that ought
to be distinguished. Where the objects themselves
are so small, one must expect the characters to
be sometimes minute: but, as far as my experience
goes, I should say that the species of Mosses are,
in general, more clearly marked and less variable
than those of Ferns.

I am very glad you like my paper on Madeira and
Teneriffe; no other of my scientific memoirs were
written with so much pleasure as that: it was a
labour of love, and is likely therefore to be as good
as anything I am capable of.

I am gratified by your saying that I give you
encouragement and stimulus to attend to botany :
it is a great pleasure to me to answer your questions,
and give you any assistance I can in that way, and I
hope you will write to me whenever you find
leisure and inclination for a little botanical chat. I
am sure my collection is largely indebted to you,
especially for those splendid Indian Ferns. Mauritius
seems indeed to be very rich in Ferns, and to be
altogether a very interesting island, botanically, but
plants from thence are very uncommon, I think,
in collections. There is a notice of its botany, by
Gardner, in the 3rd vol. of the *London Journal of
Botany*, and a more detailed account by Gaudi-
chaud, who says that it has 256 genera, and 700
species of flowering plants, and 81 species of Ferns.
Many of its Ferns are identical with East Indian
species. I have a very few Ferns from thence, given
me by Harvey, (whence he got them I do not know),
and those are the only Mauritian plants I have
seen.

Poor Miss Parker,* I hope she will soon recover
her health and eyesight, it is a most sad privation
for her, and very sad that it should have been so
caused.

I will say little of the Peace : it would be
premature to find fault with it before we know its
conditions, but I shall be agreeably surprised if it
proves satisfactory, even looking at it (as I do) in an
English point of view. It brings comfort to a vast
number of anxious hearts, and I am not disposed

1856.

* A dear friend and governess in his wife's family.

1856. to depreciate its blessings, but yet I cannot help
wishing that we could have given the Muscovites
one thorough good beating while we were
about it.

Did you see in the *Examiner* an article on the
additions made by Lord Dalhousie to our Indian
dominions ?   It is a formidable account of the
territories and population *annexed* within so few
years : and I must say it makes one feel rather shy
of abusing the Russians for their ambition.   The
article (by Mr. Crawford, I fancy) is. an able and
vigorous one.

I hope your little heroes are quite well.
Give them my love, as well as to your husband, and
to Charles and Mary and Susan.

Ever your affectionate Brother,
C. J. F. Bunbury.

Mildenhall,
April 8th, 1856.

My Dear Mary,

Many thanks for your letter, with the
information about my brother Hanmer.   I am very
glad you like my Madeira paper : it was written
with more pleasure than any other of my scientific
writings, and is likely to have more of the freshness
of actual observation about it.

I want to see Murchison's new geological map of
Europe, which must contain a great deal of
information ; I dare say you have already seen it.
I have been reading nothing new in geology lately,
except Austen's very theoretical and ingenious

paper on the extension of the coal under the South- 1856.
eastern counties of England, much of which I find
very difficult.

By the way, I do not think I have ever thanked
Lyell for sending me the report of his lecture on the
Temple of Serapis, which interested me very much ;
the new matter was curious and valuable.

Mrs. Pellew and her son Arthur are with us.

The dogs are in noisy health and spirits. Skye
has been created a Count—*Count Bowwowsky.*

We are still in the dark about Henry's probable
movements, or rather those of his division and
regiment. It must take a long time to remove the
whole of the allied armies from the Crimea, and
then we do not know where the regiments may be
quartered : but if there is no immediate prospect of
another war, I should hope Henry would go on half-
pay, and not to the Colonies. Have you seen
Edward lately ?

Much love to Lyell and to Susan.

> Ever your very affectionate Brother,
>
> C. J. F. BUNBURY.

> Mildenhall.
> May 12th, 1856.

My Dear Lyell,

I have been quite surprised on looking back
to your last letter, to see that its date is so far back
as April 30 ; I did not know I had left it so long
unanswered. It interested me much nevertheless.
I am very glad you are trying experiments on the

1856. power of seeds to endure salt water. Of the Carices I think but few are enumerated among the plants which have a very wide range : Carex cæspitosa is named by Brown as one of those common to Europe and Australia ; and Carex Pseudo-Cyperus is (I believe) common to Europe and South America ; but I do not know of any common to the Cape of Good Hope and other countries ; and what is rather remarkable, the North-American species, which are very numerous are almost all (I believe) different from the European. Darwin perhaps would say that they are readily modified by climate or other causes, and therefore are not recognized as the same species. But many other plants of the same natural order (Cyperaceæ) are very widely diffused. Thus, Scirpus lacustris is common to Europe, North America, Cape of Good Hope, and New South Wales; S. maritimus, and fluitans to Europe, the Cape, and New South Wales; S. triqueter to the first and last of these countries; Cladium Mariscus to Europe, Jamaica and New Holland. It is most probable that these plants have the same facilities for migration as the Carices, though Darwin has ascertained that the power of enduring salt water sometimes varies greatly in plants of the same family. There is no part of natural history so interesting to me as the geographical part, comprehending not merely the actual ranges of plants and animals, but the theories of dispersion and variation and linking itself on to all the questions about species, &c. The study of fossil plants might

have an important bearing upon this branch of 1856.
science, if one could trust to the data which they
afford.

Wishing to make out,—with a view particularly
to the fossil floras of the tertiary and post-tertiary
ages,—how far the venation of leaves (of dicotyledons) can be trusted as indicating affinities, I
have begun to go regularly through the principle
genera of recent dicots in my herbarium, examining
and comparing the leaves minutely, species by species,
making notes of them as I go on. I have as yet
only gone through the Myricaceæ (Myrica and
Comptonia) so it would be premature to draw any
conclusions.—By the way (though it does not bear
upon our main subject), you no doubt remember our
all remarking that the Myrica *Faya* wants the
aromatic smell which is so characteristic of the
other Myricas and of the Comptonia. I have been
rather surprised to find that it has nevertheless the
same sort of glands on its leaves that they have, and
not in small quantity. You surprise me by saying
that Hooker was one of the party who "ran a
tilt" against species. In all his writings, even in the
most recent, his "Flora Indica," he distinctly and
explicitly maintains the reality of species, though he
holds (and I have no doubt he is right) that a large
proportion of the species admitted in our systematic
works are *not* valid. Darwin goes much further in
his belief of the variability of species, than I am
disposed to do, but even he, I imagine, would not
assert an *unlimited* range of variation : he would
hardly, I conceive, maintain that a Moss may be

1856. modified into a Magnolia, or an oyster into an alderman ; though he seems to hold that all the different forms of each natural group may have sprung from an original stock, even (for instance) that the Ericas of Europe and of the Cape may have had a common origin : which I am not disposed to believe. The Primrose and Cowslip are certainly a remarkable instance of variation, as Henslow seems to have ascertained that they may both be raised from seed of one plant, but do not forget that Linnæus considered them as one and the same species, from their characters alone, without having any such experiment to rely upon. The fact you mention about the rapid spread of an introduced species of freshwater shell over Madeira, is very curious. I should not have supposed those creatures to be such good travellers.

The analogy of the Miocene Flora of Europe to the existing Flora of N. America, has often been remarked, and is certainly very striking. It is shown particularly in the existence of tertiary fossil species of Comptonia, Taxodium (*very* like the recent deciduous Cypress) Liquidamber, Juglans or Carya, and a Vine much resembling the American Vitis vulpina. Professor Braun even doubts, as he told me, whether the fossil Taxodium be specifically different from the exisiting American one. Smilax, which you mention, is not a peculiarly American form. But, in calling all these Miocene, I follow Unger, without being at all sure whether the geological age of the different lignite-deposits in Germany, Styria, and Croatia, &c., has been satisfactorily made out. It rather appears as if 1856. Unger assumed that all " brown-coal" must be Miocene. I agree with you in thinking Heer an uncommonly clever man, and if ever I live to go to Switzerland again, I will make an effort to see him. I should like to send him a copy of my Linnæan Paper on Madeira, if I knew of an opportunity. I am really afraid something must have happened to my Italian correspondent, De Zigno, not having heard a word of or from him in reference to the drawings which I sent him early in February.

I am reading Dr. Sandwith's book with very great interest. What a woful picture of the state of the Turkish provinces! The " sick man " does seem to be very sick indeed. I am very glad General Williams' services are acknowledged in such a marked way by our Government, and pleased also with the way in which Count Walewskis' atrocious attack on Belgium was noticed by Lord Palmerston.

Much love to Mary ; I hope we shall meet early in June.

Ever affectionately yours,
C. J. F. B.

Mildenhall,
May 20th, 1856.

My Dear Susan,

I have long intended to write to you to thank you for the Papers of the Sunday League, and to assure you that I am quite of your way of thinking on that subject. (Thanks to the

1856. Scotch Members with your friend the Lord Advocate
at their head, and thanks also to the cowardice
or indifference of Lord Palmerston), they will
probably be for pushing further, and if we do not
take care, we shall have the Sunday completely
puritanical. I am not fond as a general rule of
leagues or associations for carrying political or
social questions, because one is exceedingly apt,
sooner or later, to find one's self implicated in
proceedings that one cannot approve; but in this
case the combination on the other side is so
formidable, and appears so much more than it is,
from the number who are afraid to let their opinions
be known, that I really think it becomes a duty to
give one's name, if one can give nothing more. I
cannot promise a large subscription, because I am
labouring under the complaint of impecuniosity;
but what I can give, I will. I should not much
wonder if my Father were to give you his name, for
he is very indignant against the sabbatarians, and
he asked to see the prospectus of the League, which
we accordingly sent him.

We enjoyed very much the visit we had from Mr.
and Mrs. Horner and Joanna, and were very glad
indeed to have your mother with us a week longer,
and I hope she was really the better for the stay
here. She will perhaps have told you of our
reading " Westward Ho!" I wonder whether she
will have the curiosity to go on with it. I have
finished it since, and think it a very admirable
novel; I do not deny that there are parts which are
heavy, but altogether it interested me very much,

and I think it powerfully and beautifully written, 1855.
and very noble in feeling. I have been reading
the same man's (Kingsley's) sermons, and am
delighted with some of them. I am just finishing
Dr. Sandwith's book on Kars, which is indeed
most interesting; but what a dreadful picture of
the state of the Turkish provinces and of the
corruption of Turkish officials. The "sick man"
does appear to be very sick indeed, and one cannot
help having painful doubts, whether even the
Sultan's apparently wise and ample measure of
reform will have any practical effect. Yet the
conduct of the Turkish private soldiers, and what
we hear of the peasantry, seems to show that there
is good matter at the foundation. What is peculiar
and puzzling in the case is that the evil does not lie
(as in Spain at the time of Napoleon's invasion) in a
degenerate and corrupted *privileged class*, but every
man in turn seems to become corrupt, as soon as he
is raised into any place of power or emolument. It is
delightful to read of such men as General Williams
and his comrades; they do honour to their country,
and show what Englishmen can still do when
their energies are not crushed by a vicious system.
I daresay you are very angry with the peace. For
my part, my opinion of it is something like what Lord
Derby expressed, that it is a peace to be tolerably
well satisfied with, but not a subject for exultation,
or for any great rejoicing; that we might have had a
worse treaty, and we might have had a better. I
rather hope than expect that our Government may
be better prepared for war next time, and how soon

1856. that *next time* will come, who knows? The excessively ungracious and offensive, not to say unjust way in which the authorities have begun to deal with the officers the moment peace is concluded does not look well.

Addio dear Susan; I look forward with great pleasure to meetings in London before very long, and am ever, with much love to all the family party at 17, 14 and 53.

Your very affectionate Brother,

C. J. F. BUNBURY.

Mildenhall,
May 29th, 1856.

My Dear Lyell,

Many thanks for sending me Heer's letter, which I have read with great interest. It is very clever and ingenious, showing, like his printed essay, a very acute mind, much addicted to bold speculation and theory. You as a geologist must deal with his speculations concerning the junction of Europe with America, and its separation from Asia in the Miocene age, its climate, &c., all which appears to me abundantly bold. I agree with you in thinking that there is some confusion or indistinctness in his reasoning in that first part of the letter, which appears to be based on an implied assumption *(not* expressed) that *representative* species are likely to have proceeded from a common stock, *i. e.* that species are not fixed and constant creations. It is quite true that we are much in the dark as to the limits of variability of species : but

there is no reasoning clearly or understanding one 1856.
another in these matters, without some previous
explanation of what one means by species. I have
already said, in another letter, that I can see
nothing in the flora of the Atlantic Islands, to
require or justify the supposition of a former connection with America. And with respect to the
Miocene flora, it strikes me that, before we allow
much weight to Mr. Heer's reasoning on this head,
we ought to know something of the Miocene flora
of America.

If the middle tertiary flora of the U. S. should
turn out to be materially different from that of a
nearly corresponding age in Europe, it appears to
me that this would be a stronger argument *against*
Heer's theory than any he has advanced *for* it.
It is very odd that Heer should say that "the genus
Platanus is *entirely wanting* in actual Europe:" he
seems to forget the Oriental Plane which I believe
is undoubtedly a native of Greece and Turkey,
though it is said to have been introduced into
Sicily.

The tertiary species of Plane which Heer
mentions, seems by his account to differ so slightly
from the recent American one, that I should be
inclined to think it may really be a mere variety of
that; the differences seem no more than what one
may suppose to occur within the limits of one
species. As to the Taxodium,—Professor Braun
of Berlin told me that he really believed the fossil
European plant to be identical with the existing
American one.

1856.   These therefore would seem to be instances
decidedly favourable to Heer's doctrine.

Whatever Hooker may be, I am certainly not
in all cases sceptical as to the determination of
fossil dicotyledons.   In such cases as the tertiary
species of Taxodium, Platanus, Liquidamber,
Hornbeam, Birch, and perhaps a few others, where
the fruits and other parts have been found in
company (though not in actual connection) with the
leaves, I am quite ready to believe in the identifi-
cations, also where the leaves have a very peculiar
and strongly marked character, as Liriodendron,
and perhaps Comptonia.   But where the leaves
alone are found, and those of a very ordinary
character, I cannot help being very sceptical as to
the power of determining them.   I gather from
Heer's letter that he relies mainly on the veins of
the leaves, and I am quite aware that in some
instances, those afford good characters, but I wish
he had given some of his reasons for thinking that
they do so *generally*,--or some examples of their
value.   As I mentioned in my last letter, I am
working at this subject, examining minutely the
venation of all the dicotyledon trees and shrubs I
can get at, group by group; and I must say, that
as yet, I have found no reason to think that this
part of the structure affords good generic characters.
I find it neither constant in the same natural
genus (such as Quercus) nor always distinguishable
in very different genera.

Now to come to the subject of your letter received
this morning : I have carefully drawn and described

the greater number of the forms of leaves that I 1856.
can make out in your S. Jorge collection : but have
not finished the whole, nor have I yet written anything on the generalities of the subject.

I am happy to say my little nephew and niece* are
both so much better, that we are now almost
relieved from anxiety about them.

With much love from both of us to dear Mary, I
am ever,

Yours affectionately,

C. J. F. BUNBURY.

Mildenhall,
May 31st, 1856.

My Dear Katharine,

I hope to see you very soon now, so I will
not attempt to write fully this time and I will bring
you back your list of Ferns, with many thanks,
having noted from it the names of all the Indian
ones that have numbers to them.

I most heartily agree with you as to the pleasure
of exploring a rich botanical country ; often I
think over my days in Brazil, at the Cape, or in
Madeira and Teneriffe ; and often I long to be able
to visit the West Indies, or the Indian Islands. I
have gone through that parcel of Himalayan plants
which you so kindly gave me, and transferred most
of them to their proper places in my collection ;
several are very interesting to me. I hope we shall
have some good botanical talks while I am in town.

Dear Cecilia and her two little ones have been
staying with us some time ; she is always

---

* Emily and Henry, children of Col. Henry Bunbury.

1856. charmingly good and pleasant; and her children are to me very engaging; the dear little boy was alarmingly ill for some days, but is now thank God, quite convalescent; and little Emily, who is quite my delight, and who has also been delicate, is going on very well.   Both are very fond of Aunt Fanny, and I think it will be rather a pang to her to part from them.   I hope your dear little men are quite well.

It is very fortunate that the fireworks and illuminations were so successfully managed, without, as it appears, any accident or disturbance at all. The weather, here at least, has been very cold and disagreeable for some days past, so that we have been glad to come back to fires and winter clothing : I hope it may be pleasanter on the 10th.

I will write no more just now, as I hope to see you all so soon.

Ever your very affectionate Brother,
C. J. F. BUNBURY.

JOURNAL.

June 4th.

Arrived in London.

June 5th.

Visited, but hastily, the exhibition of the Royal Academy.   Exhibition of French pictures :—two admirable Sea-pieces by Gudin, of which one, a View on the coast near Aberdeen, with the moon rising, and the sea rippling softly in on a level sandy 1856. shore, very beautiful. "Napoleon crossing the Alps," by Delaroche, which I knew before from prints, I like much. Two pictures by Anatole Beaulieu,—"A House Scene in Algiers," and "A Serenade in the Carnival at Venice,"—very fine in colouring, like Etty. There are several other clever and pleasing little pictures of domestic or *genre* subjects. Judging from this exhibition, the French artists would appear to be like our own, much more successful in landscape, animals, and "genre" picture than in historical or political subjects. The scriptural pictures here appear to me very feeble : the "Battle of the Alma" unsatisfactory; the great picture of Ney on "The Retreat from Moscow," unpleasing and not impressive.

At the Athenæum, met Charles Napier,* lately returned from the Crimea. In his way back he stopped at Athens : and he speaks strongly of the miserable and hopeless government of that country. The Greeks themselves, he thinks, a people from whom (with all their faults) much may yet be expected : but the government is at once feeble and oppressive, corrupt, shabby, and showing in all its fiscal measures the most ruinous ignorance of the first principles of political economy. The country is in a most lawless state, overrun by strong bands of robbers, who are chiefly disbanded soldiers, and men who have taken an active part in the War of Independence, and whose pensions, allowed for their services in that war, were stopped by the

* Son of Captain Henry Napier.

1856. present Government. Even in the immediate neighbourhood of Athens, the robbers are supreme; they robbed a Government convoy very near the city; they carried off a French officer to the mountains, and the Greek Government had to pay his ransom. At the same time Charles Napier speaks highly of the spirit and cleverness of the Greeks. At Constantinople he was much struck by the laziness and helplessness of the Turks.

———

June 6th.

Dinner party at Charles Lyell's. Mrs. Jameson, Mr. and Mrs. Shaw, from Boston, the Joseph Hookers and Erasmus Darwin. The fireworks* and illuminations on the 29th were talked of, and Mrs. Shaw said nothing had struck her so much on that occasion as the vast multitude of people and their excellent behaviour. She admired also the police of London.--Hooker spoke of the great utility of our exploring and surveying expeditions in recent times as the best schools for our seamen and officers, and best adapted for exercising all their most important and valuable qualities, as well scientific knowledge as practical qualities. He is anxious that the search for the relics of Franklin's expedition should not be given up till everything be ascertained that possibly can be. Mrs. Jameson said that she, like we, had been much disappointed by the Correggios at Dresden. In her visit to the Continent last year, she was much struck by the superiority of the Piedmontese to the other Italians.

* On account of the Peace.

Afterwards a large evening party: I had much 1856.
pleasant talk, but learned nothing particular.

———

At the Athenæum: looked through Wollaston's
fine book on the "Insects of Madeira." The
newspapers announce the death of old Bishop
Monk.*

Again to the Royal Academy,—"The Scape
Goat," by Hunt,† a singular picture, with great
merits, I think; the feeble, exhausted hopeless look
of the animal, the dim glazed eye, the tottering step,
the salt crust crackling under his feet, the sunset
glow on the savage mountains in the background,
all appear very truė to nature, but it is scarcely a
pleasing picture. "Chatterton," by Wallis (a new
name), a finely-coloured and impressive picture.
"The Wreck Abandoned," (Stanfield): fine, but I
think there is much truth in the criticism I have
heard,—that the waves in the picture are the broken
dashing waves, seen near a coast, whereas by the
story, they ought to be the waves of the open ocean,
and consequently of quite a different character.
Roberts's views in Venice are charming. There is
a picture of "Roman Women in a Balcony during
the Carnival," by Lehmann, which pleases me
much: as also one or two pictures of Spanish
figures, by Phillips.

The portrait of Dr. Sandwith is interesting.

Poor. Daniel Sharpe‡ was buried to-day. His

---

* Bishop of Gloucester.   † Holman Hunt.   ‡ Nephew of Samuel Rogers.

1856. death is a serious loss to geology. The office of President of the Geological Society is offered to John Phillips.

June 8th.

Visited my Uncle and Aunt at Clapham. Sir William is convinced that the Americans will go to war with us, though he does not suppose that the war will be popular in America any more than here. He thinks that the great number of wild, restless, reckless, adventurous young men in America, with nothing to lose and much to gain, will hurry on a war ; and that the Government will favour it, as the means of keeping off a civil war between the Free State and Slavery parties. As to our state of preparation, he said—we have a very fine fleet, and a great army, and the most ignorant set of commanding officers, and the most ignorant Ministers, that ever we had ! He is not apt to look at the bright side of things.

June 9th.

Went into the Vernon Gallery, and looked at several favourite pictures. At the Athenæum had a a good talk with Joseph Hooker. The East India Company has behaved most shabbily to him and Thompson about the "Flora Indica." He says there are no certain limits between the genera Oak and Chesnut ; that the Indian species break down all the supposed distinctions. He talked of the extreme variability of the Conifers, especially of the Junipers, and agreed with me in thinking that there is no genus in which it is more difficult to fix the

limits of species than in Juniperus. He gave me a 1856.
curious instance of the variability of the Deodar
from seed. He lately saw at Bury Hill* some
Deodars raised from seeds out of the *very same
cone* from which the largest Deodar at Kew was
raised, and yet totally unlike it in habit. He thinks
it an unsettled question whether the Pinus Pumilio
be distinct from Pinus sylvestris. On the descent
of the Grimsel, he thought he could trace a series of
variations from Pumilio into sylvestris. He is
sceptical, as I am, as to those determinations of
fossil dicotyledonous leaves, in which the continental
geologists have lately been so active and confident;
that is, as to the *generic* determinations of those
which are *not* existing species. He doubts whether
the venation will generally afford valid generic
characters. He told me of some truly silicified stems
believed to be of Arundo Donax, and of a very recent
age, which Robert Brown has lately received from
Egypt; also of silicified wood believed to be of
Banksia, found in great quantities in Tasmania, in
a different locality from the Coniferous wood. The
wood of Banksia is characterised (like that of the
Cape Proteaceæ) by large and wavy medullary
rays.

Went with Fanny to the British Institution and
saw the exhibition of pictures by the old masters—a
very good one. Several fine Vandycks, among
which one of Sir Kenelm Digby and his wife is
remarkable. A large picture by Rubens, of Villiers,
Duke of Buckingham, an odd composition, crowded

* Near Dorking.

1856. with allegorical figures, in particular, a fat, naked
nymph (finely painted however) who lies sprawling
directly under the hoofs of his horse, and looks afraid
of being trodden on, has a very odd effect. A
singular picture by Giov. Bellini of the "Gods
Feasting on the Earth," engraved in Agincourt.
"Venus Wounded," by Pagi (a Genoese artist),
very pretty. A very attractive portrait of Georgiana,
Duchess of Devonshire, by Hoppner; and one of
Lady Hamilton, by Romney.

June 10th.

A beautiful day. The fiftieth anniversary of Mr.
and Mrs. Horner's wedding day. Grand family
merry making.

Lyell thinks (and so does Mr. Pulsky, whom I saw
the other evening) that we shall not have war
with America immediately, but that much will
depend on the Presidential election in November
next; that if Pierce should be re-elected, or
another of the democratic and pro-slavery party
be elected, the two nations would certainly be at
war before long.

June 11th.

A most beautiful day. Went with Mrs. Horner,
Joanna and Charles Pertz, to the Botanical
Gardens in the Regents Park. The grounds are
remarkably pretty, and the gay assemblage of well
dressed ladies had a very agreeable effect. The
exhibition of Rhododendrons and Azaleas astonished

me by its extent and beauty; a large space of 1856.
ground, protected from the sun by canvas, was
covered as thickly as possible with dense masses
of blossom, of every shade of lilac, pink, white,
yellow and orange, these last two colours being
supplied by the Azaleas. The effect was
wonderfully beautiful.

June 12th.

A stormy day. Read some of Dugald Stewart's
Dissertation. Looked into the National Gallery,
and saw some of the pictures newly added to it;
Titian's "Christ and Magdalene," from Mr. Rogers'
collection; Bassano's "Good Samaritan," from the
same; Rubens's sketch of "The Triumph of Julius
Cæsar," from the same. This is a remarkable
work, professedly copied from part of Mantegna's
great composition, but so much altered that it must
be called a very free translation, strongly imbued
with the characteristics of Rubens's manner. The
great picture of "The Adoration of the Magi," by
Paul Veronese, lately bought, does not appear to
me a fine specimen of that master.

Met Hugh Adair, and congratulated him on his
approaching marriage. Heard from him of poor
Mr. Eagle's death. At the Athenæum, saw John
Moore, Woronzow Grieg,* and Mr. Boxall. John
Phillips having declined to undertake the Presidency
of the Geological Society, Colonel Portlock is to be
President.

---

* A Son of Mrs. Somerville by her first Husband.

Much rain.   Read some more of Dugald Stewart.
Paid some visits with Fanny, and afterwards went
with her and Mary and Joanna to see Stafford House.
It is really a palace; very magnificent; the great
central hall and the picture gallery, as fine as
almost anything I remember to have seen; but the
pictures rather disappointed me.   There is a charming Landseer.

At the Athenæum; met Sir Edward Ryan and
Mr. Bentham; the latter very much pleased
at the prospect of the establishment of the Linnæan
Society in Burlington House, which he thinks will
be very advantageous to it.   Sir Edward Ryan
thinks the difficulties of the American question
very great, though to me the last news does not
appear so very unfavourable; for though the
Americans have, it is true, dismissed Mr. Crampton
and three of our Consuls, yet the despatch of their
Secretary of State appears to be concilatory, and to
afford a prospect of a friendly settlement of the
dispute.

Showery.   Went to the Geological Society,
where Rupert Jones showed me some curious
specimens of fossil fruits and seeds lately received
from India; also the fine palaeozoic fossils from
South Africa; which he has been employed in
arranging.   Looked at Murchison's beautiful geological map of Europe.   Then to the Athenæum,
and spent some time in reading Sir Edward Cole-

brooke's Journal of his "Two Visits to the Crimea," 1856.
which interested me a good deal.

***

(Sunday). Paid a round of visits with Fanny;
saw Norah Bruce, Miss Phillips and Lady Bell.
The latter told us a good story, which she said Sir
James Mackintosh used to be fond of telling. Once
on a voyage to India, a married lady happened to
fall overboard. Her husband cried out—"Oh save
her! save her! I will *give five pounds* to the man
who saves my wife!" The lady was saved. Some
time afterwards she eloped with another man. The
husband brought an action for damages against her
lover, but unluckily for him, it came out that he had,
on the occasion when she was in danger, valued her
at £5; and the damages were assessed accordingly.
By the way, this story is a curious illustration of the
commercial light in which conjugal infidelity is viewed
in England.

***

Paid some visits with Fanny; saw Mr. Hallam,
looking much changed and broken, sadly infirm and
deaf. We afterwards went to Cremorne Gardens, a
sort of new Vauxhall, by the river-side beyond
Chelsea; a pretty scene. There is a show of
Rhododendrons here, very beautiful, but not equal
to that in the Regents Park.

***

It appears our Government have decided *not*
to retaliate for the dismissal of Mr. Crampton by

1856. dismissing Mr. Dallas, but to carry on negociations
with him. I think they have decided wisely. Mr.
Marcy's dispatch is said to be temperate and
conciliatory. A pleasant evening party at the
Horners. I had some talk with Mrs. Douglas
Galton, Miss Richardson, Mr. Babbage, Mr. Boxall,
&c., &c.

June 18th.

With Mary to the flower-show in the Botanic
Gardens in the Regents Park. The beauty of the
day and the multitude of gaily dressed ladies, made
it a very pleasing scene. The show of flowers was
very fine ; Cape Heaths, Roses, Pelargoniums,
Calceolarias, in great variety and beauty. A
splendid display of Orchideous plants, in particular
most beautiful specimens of Phalænopsis, and
magnificent Cattleyas. Lælia cinnabarina, striking
from the singularity of its colour, which however
is rather that of red lead than of Cinnabar.
There were some fine specimens of Ferns, in particular Cibotium Schiedii. A magnificent Allamanda
with its great golden flowers, and " Dipladenia,"
(Echites) crassinda, with large blossoms of the
finest rose colour, these flowers in both instances
relieved by dark green leaves, extremely beautiful.
An astonishingly large specimen of Roella ciliata,
with a very great number of flowers. Dined at the
Geological Club. Present, Colonel Portlock,
(the President) Murchison, Lyell, Mr. Horner, Dr.
Percy, Mr. Warrington Smyth, Colonel Tremenheere
(as a guest), Mr. Mylne, and one or two others.

There was some good talk. Murchison extolled the 1856. talents of a young sculptor named Noble, who was the son of a poor mechanic at Hackness in York- shire, brought forward by Sir John Johnston, and educated at his expense. He (Murchison) thinks him superior to Chantrey. This led to talk of Chantrey ; Murchison mentioned the piece of ornamental furniture wrought by that sculptor in his young days, which Mr. Rogers had, and which he was fond of showing, and telling the story of ; and he remarked how singular it was that an article, interesting on account both of Rogers and of Chantrey, should have been sold at Rogers's sale for only ten guineas. At the same sale, a small picture by Leslie, (" The Duchess and Sancho,") for which Rogers had paid £70, sold for £1180. Mr. Leslie him- self told Murchison it was a most extravagant price. It is said the Manchester men are the great buyers of English pictures. They avow that they cannot under- stand the old Italian pictures, but they will give almost any price for good specimens of the English school. Methinks they are very wise to buy what they can understand and relish.

At the evening meeting of the Geological Society, Colonel Portlock was formally elected President. Several short papers were read ; the most interesting to me was a notice of the geology of the country about Varna, by Captain Spratt. He noticed those curious pillars near Aladyn, which excited so much curiosity, when our army was encamped there in '54, and which were taken by the officers generally for ancient works of art. He describes them as com-

1856. posed of nummulitic limestone, and considers them as concretions exposed by the natural wasting of the rock; but their almost regular cylindrical form is not accounted for. I was induced to come away before the end of Prestwich's paper, and so lost what I have since heard was a very good discussion.

June 19th.

To the Exhibition of Water-colour Paintings, where, as usual, many beautiful landscapes. Lewis's " Frank Encampment in the Desert of Mount Sinai," is a very remarkable and interesting, and I should think a true, picture.

Spent much of the day in quiet reading at the Athenæum; studied Humboldt's and Bonpland's " Plantes Equinoxiales," and Royle's " Himalayan Botany," and read a good deal of the seventh volume of " Moore's Memoirs," which is very entertaining, and less frivolous than some of the earlier volumes.

Went with Mary and Susan to an evening party at Sir Charles Eastlake's where there were several handsome women, but few people that I knew. I had some talk with Mrs. Jameson; she remarked (what indeed I have heard others also remark) the increased and increasing splendour and extravagance of women's dress in these times. The amplitude of the petticoats is becoming perfectly enormous. The Countess Appony (wife of the Austrian Ambassador) at this party, displayed a most conspicuous example of this.

All Addison's ridicule of the hoop petticoats of

his time might well be reproduced now. Indeed 1856. they say that actual metallic hoops are coming into use.

I had some geological talk with Murchison, who gave me an account of the discussion at the Geological Society last night on the nummulitic formation. He contends (and I think with great reason) that the vast formation of nummulitic limestone in Asia and South Europe is only *in part* represented by the beds which Prestwich and Lyell consider as its equivalents in middle Europe, and that it corresponds to the Lower Eocene *as well* as the Middle Eocene of England, &c. There are some very fine pictures at Sir Charles Eastlake's especially two Venetian ones.

——————

June 20th.

Charles Darwin came in at breakfast time, and I had an interesting talk with him about *species*, and the various questions connected with their origin, distribution and diffusion. All this—all connected with the geography of natural history, in the widest sense, is to me the most interesting part of the science, and it is that to which Darwin has long devoted himself. I was very glad to find that there is some prospect of his publishing his views on the subject. He spoke with great admiration of Alphonse De Candolle's new work on botanical geography, though he said that Joseph Hooker does not appear to think so highly of it. He is sceptical about the *Atlantis*, of which according to Bory, E. Forbes and Heer, the Atlantic Islands are the

1856. remains ; and thinks generally that the theory of the migrations of plants and animals by land since submerged, has been carried too far.   He thinks that much yet remains to be learned with respect to the means of transport of plants, and mentioned in particular some observations which led him to believe that the seed of plants might sometimes be transported in earth enclosed amidst the roots of floated trees.   He believes also that the agency of birds in the transport of seed has been underrated by De Candolle.   He says he has ascertained by careful experiment that seeds of West Indian plants cast up by the sea on the Coast of the Azores, have germinated.   He spoke of his theory (which he had before mentioned to me by letter) that an interchange of plants, to a certain degree, may have taken place between Europe and the Cape of Good Hope, during the glacial period, when a much colder climate than the present existed beyond certain parallels of latitude, in both hemispheres. In accordance with his doctrine of the mutability of species, he supposes that the representative species in either hemisphere (*e. g.* the species of Dianthus at the Cape, and those of Gladiolus in Europe), may be modified forms originating from the opposite hemisphere in which their respective genera have their head quarters.   His chief difficulty is, that there are so many more representatives of northern forms at the Cape than representatives of southern forms in Europe ; as if the migration had taken place chiefly from north to south.

To Kew gardens. I was sorry to find that the Tree Ferns which used to be in such beauty in the great Palm House, are nearly all dead. Some of the Palms are glorious, especially the Caryota urens, and two species of Cocos from Brazil. Met with Bentham at the railway station, and came back to town in the same carriage with him ; had much botanical talk. He has been much engaged with the Brazilian Leguminosæ, working them up for Martius's Brazilian Flora. He says that species of *Desmodium*, even more than other Leguminosæ, are difficult to determine without their fruit, and there has been much confusion about the South American ones : several of which have a considerable range in latitude. He remarked how much confusion and perplexity has been caused by the modern rage for multiplying genera on slight grounds, that no one will now submit to be guided by authority in such matters, and every young botanist seeks to distinguish himself by making fresh subdivisions. Speaking of the *Loganiaceæ* (on which he has lately written an excellent paper), he remarked that they are not a very natural group, yet on the whole he thinks the several genera more allied to one another than to any other family. The only absolute distinction between Loganiaceæ and Rubiaceæ, he says, consists in the ovary, free in the former, adherent in the latter, and this *almost* breaks down in such cases as Houstonia ; yet it is advisable to keep the two orders separate. Gelsemium, he says, is almost a Manettia with a free ovary. Other

1856. genera again though with so little to separate them technically from Rubiaceæ, have really much more affinity to very different orders ; for the affinities of Loganiaceæ branch off in many and various directions.

June 22nd.

Went to church at St. Mary's, Bryanston Square, and heard a good sermon. Visited the Youngs. Mr. Young thinks that the immediate risk of a war with America is over, but that Central America will long continue to be a sore and difficult point.

Walked in the Zoological Gardens .with Mary Lyell. The animals in fine condition. Saw the white bear bathing. The cry of the spotted hyæna is peculiarly horrid and hideous. The Lyells and Edward dined with us ; much pleasant talk. Lyell thinks that the question between the free and slave States of America is coming to a crisis, and that it is particularly desirable that they should be left undisturbed to work it out ; that a war with England at this time is the very thing that would best suit the purposes of the pro-slavery party. He thinks that this critical struggle will in the end prove very advantageous to America, but in the meanwhile will for a long time keep it too much engaged to interfere with other countries.

June 24th.

To the Museum of Practical Geology, and saw Salter. He showed me the fine collection of fossil leaves from Alum Bay, &c., which De la Harpe of 1856. Lausanne examined when he was in England last winter, and to which he affixed names.

We went, a large party (four Horners, the Charles Lyells, Minnie Napier, ourselves, and Edward) to the Princess's Theatre, and saw the Winter's Tale. The scenery, costumes, decorations, and in short the whole "getting up" of the play are most beautiful. The costumes classically correct, and well-adapted to give an idea of the grace and beauty of all that belonged to daily life among the ancient Greeks. Some of the scenes looked like realizations of ancient vase-paintings or Pompeian pictures. A Pyrrhic dance in the first act, the trial scene in the theatre of Syracuse, and a Bacchanalian dance, were peculiarly beautiful. In short, it was one of the finest *spectacles* I have ever seen. The acting not remarkable : though Mrs. Kean was good in Hermione, and Perdita was very prettily acted by Miss Leclerc (I think).—Harley also very good in Autolycus, Mr. Kean as Leontes, did not please me.

————

June 25th.

Peculiarly oppressive sultry weather, seeming to threaten thunder, but none came. Went with Susan to the Botanical Gardens in the Regents Park, and met the Lyells. Had some talk with Charles Lyell on the extensive question of *species*. We dined with the Youngs; met Dr. Bence Jones, met Mr. and Mrs. Paget (*he* is one of the Pagets of Yarmouth), Fanny Pellew, William Winthrop, &c.

June 26th.

Weather exceedingly sultry.   Mr. Lacaita came
here to luncheon ; he is a very agreeable man ;
Charles Napier and William Winthrop called soon
after.   At the Athenæum I read Dr. Falconer's
extremely able article in the "Annals of Natural
History," in reply to Mr. Huxley's attack on Cuvier.
Strolled in the Zoological Gardens, where the lions
were roaring gloriously.

June 27th.

Went with Fanny to Leicester Square, and saw
the panorama of the Interior of Sebastopol.   It
is very well painted, and Charles Napier, who was
with us, told us it was a very good representation of
the place and of the surrounding country.   This
was a very fine and very hot day.

June 28th.

A lovely day.   We went out to Twickenham to
dine with the Richard Napiers, and spent a very
agreeable afternoon with them.   I have always
found them delightful company and most true and
valuable friends.

June 29th.

We went to Quebec Chapel and heard Alford
preach.   I thought the language of his discourse
more excellent than the matter.

Walked in the Zoological Gardens ;   saw the fox-
headed bat or "flying fox" (the *roussette* of Buffon

—mistaken by Sir Everard Home for the vam-
pyre), a new acquisition; a smaller animal than
I had supposed; sleeps suspended, head downwards,
wrapping its flying membranes round it like a
mantle in a curious and beautiful manner ; eats (the
keeper told us) nothing but fruit.   The clouded
tiger from Assam is a very beautiful animal; lies
habitually (in the day time) on a branch, in a very
singular and apparently stiff and constrained at-
titude, looking like a stuffed animal.

June 30th,

Went with Fanny and Minnie to the Lyceum, to
see Madame Ristori in "Pia de' Tolommei."   Her
acting is magnificent ; certainly I have never seen
such fine tragic acting.   She is a very handsome
and noble looking woman, full of grace and majesty.
The play itself is not very interesting, yet the
character of Pia herself is fine, and Ristori's noble
acting interests one deeply in it.   Her personation
of lingering death from marsh fever in the last act is
almost too painful.—This play was performed in the
morning, *i. e.* from two p.m. to between four and
five, and we found it much less fatiguing than an
evening performance.

July 1st.

Went into the Museum of Practical Geology, and
looked at the fine and instructive collection of ores
and metals in various stages of preparation.   Read
some of the "Revelations of Prison Life," by Mr.
Chesterton, who was for twenty-five years governor

1856. of Coldbath Fields Prison; it apears to be the book of a sensible and humane man, and contains many curious and interesting anecdotes illustrative of the characters of the criminal population. Dined at the Athenæum; saw there Charles Lyell, George Jones (the artist), W. Hamilton, Herman Merivale, Edward Romilly, Robert Brown, and Edward.

July 2nd.

We left our kind friends of the Regents Park, and removed to the Lyells, in Harley Street. Met there Mr. and Mrs. and Miss Ticknor, and Miss Guild, Mr. Ticknor's niece; they are very lately come to England. Lyell hears from Mr. Ticknor that the American Government has not made the least preparacion for a war with England; that none of the forts along the sea-board are in a state of defence, nor the country in any way ready; so that it is evident *that* Government had no serious intention of war, and that all its bluster was for electioneering purposes. Mr. Ticknor (who himself belongs to the Whig party in America), thinks that Buchanan, the Democratic candidate, is most likely to be successful at the Presidential election.

July 3rd.

Again to the Museum of Practical Geology; saw the bust of Edward Forbes which has lately been put up there. The collections, mineralogical, metallurgical and geological, are rich and remarkably instructive. There are some interesting things

from our Colonies, especially splendid specimens of 1856.
the carbonates of copper from South Australia ; also
tin ore from Victoria (alias Port Phillip), and
sapphire from the interior of New South Wales.

Called on Sir Charles Lemon, found him looking
well, and very friendly and pleasant.   Talking of
climate, he told me that the last winter but one,
that of 1854-5, had been one of the most severe he
ever remembered in Cornwall ; that whereas ice and
snow are rarely seen in his neighbourhood, in *that*
winter there was ice strong enough to bear skaters.
Nevertheless, that severe winter was not so
destructive to the more tender trees and shrubs as
this last spring.   He told me that he had had at
Carclew the finest Abies Webbiana in England, but
it was destroyed by lightning about three years ago.
We talked about Ferns, and their great variableness
of which he said he had observed many examples in
Cornwall.   He mentioned a singular variety he had
found of Blechnum boreale, with the fronds branch-
ing out at the top into a radiating tuft, as has
sometimes been observed in the Hart's-tongue.

Speaking of the Americans, he said he had
observed that those who came to this country in
a diplomatic capacity, generally began with
very repulsive and *defying* manners, as if they
expected to be insulted, but softened by degrees,
and generally ended with becoming very fond of
this country.

1856. Called on Mr. Donne, and had a pleasant talk with him. He said he had lately seen Donaldson at Cambridge, where he was going on very prosperously, very happy, and quite in his element, which he never was as a schoolmaster, that in fact though a very good teacher, he was not at all fit for a schoolmaster, because he never in the least understood boys, nor had the least idea how to manage them. We talked of Dawson Turner, who is still living at a great age. Donne said that he had a peculiar talent for method and arrangement, and also a great talent for making use of the knowledge and talents of others.

I looked into the British Institution, and there met Esmeade, and we compared our impressions of the pictures. He abused the Canalettis, though they are fine specimens of the master ; but certainly Canaletti is apt to be hard and cold. In my notice of this exhibition, on the 9th of June, I omitted some remarkable pictures. There is a very noble portrait of a Spanish Nobleman, by Titian : and a small sketch, very brilliant and powerful, of the portrait of Paul the third, by the same. Hogarth's famous " Sigismunda," in spite of Horace Walpole's sneering criticism, appears to me a fine picture.

July 4th.

A dinner party at Charles Lyell's, Mr. and Mrs. Ticknor, Mr. and Lady Mary Labouchere, Lord Landsdowne, and Captain Murray besides ourselves

Captain Murray was the chief talker. He is very 1856.
lively and entertaining, has great spirits and a great
flow of conversation, and many good stories. He
was much in Egypt when his brother was Consul-
general there, and gave us a very animated
description of the professional story-tellers there,
reciting the Arabian Nights.

He said that no *faithful* translation of the Arabian
Nights could possibly be published in England, for
the Orientals relish no story that is not highly-
seasoned with indecency. He was at Cairo when
the Hippopotamus (the first brought to England)
was there on its way to us : and he told us that the
creature knows him perfectly whenever he goes to
the Zoological Gardens, but shows great animosity
against him. He gave an amusing account of the
wild dogs at Cairo, and speaking of the fact that
there, as at Constantinople, Rio Janeiro, &c., they
are never known to go mad : he maintained (I think
very plausibly) that canine madness is a result of
domestication : — Captain Murray, owing to his thin
and silver-white hair, looks much older than he is ;
and he told an amusing story of a young lady whom
he asked to guess his age ; she named sixty. " You
are fifteen years wrong," he said. " Which way ?"
said she.

A large and pleasant evening party, in which
I had some talk with Donne, John Moore,
Murchison, Babbage, and several ladies. I talked
with Donne about Kingsley's novels " Hypatia " and
" Westward Ho ! which he agrees with me in
admiring highly. He thinks that Froude's " History

1856. of Henry VIII " has great merits, and that there is even some foundation for his view of Henry's character, but that he has pushed all his opinions to an extreme. Froude's turn of mind, he says, is much like Kingsley's.

Murchison gave me some information about the geology of Brighton ; he thinks that the " elephant bed " is of a peculiar formation, not belonging to the ordinary northern drift ; and he says that a raised beach, of a distinct origin and different period, resembling the actual sea beach may be traced in the same part of the cliffs, to the East of Brighton.

July 5th.

We called on Mr. Boxall, and saw a beautiful portrait he has painted of Lady Eastlake, and a very pleasing unfinished picture of a young lady. He is an agreeable and interesting man, of cultivated mind, but oppressed by habitual ill health, hence somewhat hypochondriacal and fastidious.

I went with Lyell to Mr. Bowerbanks, at Highbury, near Islington, to see his collection of fossils, which is really splendid, — a wonderful collection to be made by one individual of no extraordinary wealth. He showed us a great variety of curious specimens in illustration of his theory of the origin of flint from sponges. I could not help thinking,—and Lyell afterwards owned to me that his impression was the same, that he (Bowerbank I mean) had made out a much stronger case in favour of his theory than we ever thought he could.

In particular he showed us specimens of recent 1856. sponges which had entirely filled up the interior of bivalve shells (pectens), so as to receive the impressions of the interior markings of the shell; illustrating the way in which flint occupies the interior of fossil shells; also other cases in which the sponge had filled the greater part of the shell, but left a part vacant and hollow, even as fossil shells are sometimes found, partly hollow and partly filled with flint. It appears that the sponge enters into the shell in the state of a germ, when it is very minute, and grows and spreads and encroaches till it destroys the unfortunate shell-fish. He has also abundance of specimens showing how sponges incrust shells and zoophytes, till these are sometimes quite enveloped and concealed. To such enveloping sponges he traces the origin of the common shapeless flints, or flints enclosing evidently organic bodies; even in the commonest flints he says that the structure of sponge may always be detected under the microscope. The most difficult case for his theory seems to be that of the thin extended plates of flint which often occur in the upper chalk of the South of England; but these he explains as originating in incrusting sponges spreading widely over the floor of the sea. Of the animal nature of sponges, Mr. Bowerbank has no doubt whatever, though Owen holds the contrary opinion. Besides these things Mr. Bowerbank showed us very fine specimens of the Nipadites, formerly called Cocos Parkinsoni, from Sheppey, in some of which the peculiar fibrous structure of the husk was finally

1856. preserved. Also a magnificent series of Echinoderns from the chalk, principally from Gravesend. Also a great many surprisingly fine specimens of fossil Turtles, from the London clay of Sheppey; and more fine things besides than I can recollect.

I hear Lord Shelburne's appointment (to be Under Secretary of State) much commented on. It seems to belong to the old Whig system, of making the offices of State as far as possible hereditary in certain families.

July 6th.

Sunday. A beautiful day. Mr. Gibson called, and was very entertaining and agreeable, telling many good Roman stories in his peculiarly amusing way. There is a peculiar unaffected originality about him, a vigorous simplicity that is very striking. He is as keen as ever after his *hobby* of coloured (or rather tinted) statues, and talked a good deal on the subject. He objects, however, to the manner in which Marochetti has coloured some statues, so as to have the real colouring and appearance of life; and of course much more to the coarse daubing exhibited in the Greek Court of the Crystal Palace.

Gibson's doctrine is that the tinting should be so faint and delicate, as not to have the effect of imitation of life, but to give a certain ethereal or *spiritual* look to the statue. Here is one of Gibson's Roman anecdotes. He was speaking to a Roman of the determined and unyielding courage of the Swiss, —how they would die rather than yield. The

Roman replied—" Si Signor, sono bestie feroci,— 1856.
they are wild beasts : we, who are civilized men,
know when it is necessary to run away." Another
—a Miss Hosmer, an American lady who is
studying as an artist at Rome, employed a peasant
girl for a model : and one day this girl, rising up
after a sitting, dropped a book out of her clothes.
Miss Hosmer took it up, and found that it contained
prayers *in Latin*. She asked the girl if it was *her*
book. She said it was. "But you do not understand
Latin ?" " No. But that is of no consequence, for
the Madonna and Gesu Cristo understand Latin
perfectly."

We went out to dine, and spent the afternoon at
Combe Hurst with Mr. S. Smith : and a most
delightfully pleasant afternoon and evening we
passed : I know no man more charming, more
perfectly to my taste than he is :—so thoroughly
amiable, so gentle and truly refined, so modest, so
rich in knowledge, with a most delicate taste, and
a quiet humour. His three daughters are very
agreeable. Mrs. Smith is still at Scutari, assisting
Miss Nightingale in the care of the hospitals.
Combe Hurst is a lovely spot, and we felt the quiet
and verdant beauty of its woods most refreshing
after the noise and heat of London.

_____

July 7th.

A wet day. We left London by the 4 p.m.
Brighton train, and arrived in due time at William
Napier's, at West Ashling, four miles from
Chichester.

LETTERS.

[During this interval they spent some time at Southsea, and witnessed the return of the troops from the Crimea, accompanied by Col. H. Bunbury].

Mildenhall.

August 25th. 1856.

My Dear Mr. Horner,

I thank you much for your letter of the 13th, which gave us great pleasure, and also for sending me Charles Lyell's very interesting account of his proceedings at Berlin and on the Riesengebirge. I am very glad that you are all so comfortably assembled, so safe and well at your rendezvous in the Hartz, and that you have such pleasant walks and beautiful weather; though I am afraid it has not proved quite so interesting as you anticipated, yet I should hope it will be of great benefit to your health, and the family gathering must be very agreeable. We are passing our time very quietly, regularly, and satisfactorily here, but I shall not enter into details, as Fanny's letters, especially *that* to Joanna which she has this day sent off will give you a perfect idea of "the noiseless tenor of our way."

I am much entertained by Lewes's "Life of Goethe, of which I have now read more than half; not that I much like the style, but I think it is well put together, and very pleasant reading on the whole, and it gratifies the great desire I had to know something definite of that remarkable man. It strikes me indeed that Mr. Lewes is *fully* sufficiently indulgent to his hero's love affairs, in

which he seems to have set a very bad example. 1856.
If I rightly remember what Pertz told me, Goethe's
morality in what regarded women was worse than
appears in this book ; but even by Lewes's showing
he was very much of a *male coquette.*   Still there is
much that I admire, and sympathise with in his
disposition and turn of mind especially ; and in this
I think his influence must have been particularly
beneficial to his country in his preference of outward
nature to brain-spun abstractions and dreams ; and
his powers of intellect and versatility must have
been very extraordinary.   I should be curious to
know whether Pertz has read Lewes's book, and
what he thinks of it.

I have lately had a letter from Mr. Symonds,
(our Worcestershire friend) with some account of
the British Association meeting at Cheltenham.
He says, "although smaller as regards numbers than
" the meeting at Glasgow and Liverpool, there were
" many good men present and some good Papers. We
" missed Sir C. Lyell sadly in the geological section,
" especially as regards Professor Rogers's attack on
" the Silurian nomenclature.   Sedgwick spoke with
" great eloquence and his usual fervour, and I must
" say that Sir Roderick replied with extreme temper
" and in great good taste.   We had some excellent
" local Papers, particularly from Mr. Moore of Bath,
" on the fossils of the upper lias.   He brought some
" nodules with him containing different species of
" fish, and went so far as to say that in some instances
" practice and habit enabled him to say before the
" nodule was broken what kind of fish was contained

1856. "therein.   He then produced an unbroken nodule
"and told his audience that in it lay the skeleton of
"a Pachy———" (I cannot well make out the name
in his letter); "took his hammer, cracked the
"nodule, and there lay the fish.   Nobody but those
"very near Dr. Daubeny, could hear one word of
"his address.   The Duke of Argyll and Mr. M.
"Milnes were the best speakers.   Rogers was made
"a great deal of, as also "The Queens of England"
"(Miss Strickland).   The excursions were well
"attended, but the hospitality is I hear rather
"complained of by the *Lions*.   I was introduced to
"the Arctic traveller, Dr. Rae, at the Red Lion
"dinner, and was particularly pleased with him.
"Sir W. Jardine was in the chair and Mr. M.
"Milnes, Vice;   the Lions roared loudly and
"vehemently enough to astonish the natives.   Lord
"Stanley is but a poor speaker, and I fancy the
"Section was relieved when he was called away,
"and Mr. M. Milnes took his place.   Ramsay was
"good in our geological section, and acquitted
"himself with proper dignity.   Altogether there
"was much good feeling and the great guns seemed
"determined to make themselves agreeable."

I am working at the Madeira fossil leaves and
in connection with that subject, studying the
structure and veining of recent leaves to try to find
out whether there be any characters which will
enable us to determine the families or genera of
plants by their *leaves alone;* the Swiss and some
of the German naturalists in their researches on
this class of fossils, practically assume that there

are such characters; Hooker disputes it, and I 1856.
must own that as far as I have yet gone my
researches have not much disposed me to believe in
it. I am also studying Alphonse De Candolle's
great and very important work on "Botanical
Geography," and also going on with the study of
Ferns; and I have been persuaded to give another
lecture at Bury, some time this winter. Fanny is
as usual very busy and very useful; and I am happy
to say everybody remarks how well and how fat
she is looking. In the evenings we by turns read
poetry to my niece Louisa. We have gone through
"Marmion," and are now in "The Lady of the
Lake." I always read Scott's poetry with fresh
pleasure, though it is not now the fashion to admire
it.

We agreed in relishing the fine hot weather which
prevailed during the first half of this month; our
house and the garden were particularly comfortable
then, and it was agreeable to lie on the grass, under
the trees, after the fashion of Horace. The hot
weather broke up with some of the heaviest rains that
I ever remember in England, quite like tropical
rain, and since we have had damp, lowering, cloudy
weather, sometimes cold, sometimes *muggy* and
steamy; and autumn seems to be fairly set in.

I suppose you are now at Berlin, where I have no
doubt you will pass your time very agreeably; I can
fancy you in the library, or in Leonora's drawing
room. Your stay at Paris must have been very
pleasant and instructive. I am very desirous to
have another opportunity of studying there, and

1856. hope we may be able to spend a month or so there next Spring.

We were very much pleased with Mr. and Mrs. Bowyer, who spent a day with us when he came to inspect the Union school; they are both very accomplished, agreeable people.

Pray give my very hearty love to all your party, and believe me,

Ever your affectionate Son-in-law,
C. J. F. BUNBURY.

Mildenhall,
Thursday, September 18th.

My Dear Emily,

I thank you much for your kind and interesting letter of the 14th. I can quite enter into the feelings which you so beautifully express regarding your sister; even in slight and temporary disagreements with those whom one loves, it is seldom indeed that on coolly looking back one can feel one's self to be quite blameless: and even when one is so, I can well believe that it is difficult to *feel* so, when Death the great reconciler and softener, has intervened. "The love where Death has set his seal"—is an expression that often recurs to me, there is something always solemn, almost awful in the thought of whatever is *irrevocable*.

To pass to a less serious subject: I can fancy how beautiful Abergwynant must look when the Heath is in full blossom. There is something particularly beautiful and harmonious in the combination of flowery Heath, the grey rocks, the Fern

in its rich autumn tints, and the foliage of the 1856.
Brushwood ; a combination with which I remember
to have been much delighted both in the Highlands
and about the Irish Lakes. Even here, the
Heather has been in remarkably rich and beautiful
bloom this season, and gave quite a glow to the
ground between the Bury Belt, and the " hill." It
is odd that though we have abundance of the
common Heather, and in one spot a good deal of
the beautiful cross-leaved Heath, I have never
been able to find a morsel of the deep purple kind,
Erica cinerea, here or any where in this part of
Suffolk : and I remember Mr. Eagle saying the
same. It is plentiful near Lowestoft, however. I
have once or twice found the white variety of the
common Heather here, but it is not common. I do
not recollect the Spiræa you speak of, but I am
much obliged to you for preserving it for me.—We
have beautiful autumn weather, and go out most
days to the plantations; I have been working away
at the youngest plantation here, near the Barton
Mills corner, which having been left much to its own
wicked will, the Scotch Firs spread themselves out
and interlaced their branches so as to form a perfect
jungle, which it is no easy matter to penetrate,
and in working through it (as the most vigorous
branches are generally about the elevation of my
face), I have forcibly impressed upon me the
propriety of the German term *needles*, as applied to
the leaves of the Firs. Otherwise, we seldom have
occasion here to complain of too great luxuriance of
vegetation.

1856. At Abergwynant you seem to have a good deal to do in the way of thinning.

We are both quite well. Pray give my love to my father, and believe me,

Your affectionate

C. J. F. B.

Mildenhall,
October 4th, 1856.

My Dear Mr. Horner,

This morning I have had a very agreeable letter from Lyell, from Salzburg ; they seem to have been enjoying their tour mightily, and to be charmed with the Austrian and Salzburg Alps, which indeed by all accounts, are full of glorious beauty, and of interest of all kinds. But it appears that they have given up the idea of going further into the Alps, and I suppose are moving homewards ; but I do not make out where a letter will find them. I have also lately heard from Edward, from Courmayeur at the head of the Val d'Aosta :—in great delight with Mont Blanc and its glaciers, which he had been exploring for some weeks. He says that, after all his rambles among the Alps, he has come to the conclusion that there is nothing like the group of Mont Blanc.

I am glad that noble old man, Humboldt is still flourishing. Lord Bristol, whom we saw last Thursday, is within a few weeks of the same age ; it was on his 87th birthday that we saw him when he entertained the Bury Archæological Institute at his house ; and though he has been very ill this

summer, and is rather less active then he was, his 1856.
faculties are still perfect and his manners are
courteous and agreeable as ever.

You must be pleased at the elevation of your old
friend, Dr. Tait to the Bishopric of London. I was
very glad when I saw it in the papers. I wonder who
will be Bishop of Ripon, and who will succeed poor
Buckland as Dean of Westminster ?

We are quite well, I am thankful to be able to say,
and have had a merry time with Henry, George
Napier, and Minnie Napier, and her little girl in
the house, besides our niece Louisa. Fanny's
letters will have told the particulars. A very
pleasant time it was, and it was a comfort to see
my brother so well, after his campaigns.

Sarah Napier is the most charming child I
know, and she and Louisa struck up a great
friendship and kept a fine racket in the house.
Now we have no one with us (besides Louisa) but
Louis Mallet. Unluckily he arranged the time of
his visit (by his own choice) so as to miss all our
company, and be here alone with us ; but he is
very pleasant.

Pray give my love to dear Mrs. Horner, and to
my sisters, and with Fanny's best love,

Believe me,

Ever your affectionate Son-in-law,

C. J. F. BUNBURY.

Mildenhall,
                                             October 23rd, 1856.

My Dear Lyell,

I have been for some time intending to write to you, though I have not much to tell, except that I have been exceedingly interested by your letters, and have learnt a great deal from them: and I am particularly obliged to you for your letter to me from Salzburg. You seem to have been making indeed a delightful and instructive tour, combining the enjoyment of beautiful scenery with that of geological research. I can safely say that I have learned more from your letters than I generally do from a volume of the *Quarterly Journal.* I was most especially interested by your observations on the St. Cassian or Hallstadt beds, the deep sea equivalent of the Keuper.

I presume you have satisfied yourself that the Germans are correct in their determination of the age of those rocks: and it is very curious and satisfactory to find the supposed barrenness of the Keuper age so clearly explained. It ought, as you say, to be a warning against the assumption that, because any particular beds are barren of fossils, therefore the whole age in which they were formed was barren. The intermixture of the palæozoic and secondary types in those beds is also a very remarkable fact. I sent some extracts from your letters to our friend Mr. Symonds, in Worcestershire, knowing they would interest him particularly, as he has studied the Keuper in his own neighbourhood, and he was very much pleased. He says in his answer to me,—" The more I know of Geology, 1856. "the more I am convined that Sir Charles Lyell "is our best and truest philosopher in the science, "and that the *hard lines* we have all been so apt "to draw, will at last shade away, and excepting as "local phenomena have no existence in the history "of the planet." I am extremely glad you have investigated, and to a certain degree cleared up that extraordinary and puzzling phenomenon of Barrandes *colonies*. I perfectly remembered the account of them in poor Forbes's Anniversary Address. They appeared so unaccountable, that Forbes seems to have been sceptical as to the accuracy of the observation ; but as I understand you, you are satisfied of their real position.

Your explanation, illustrated by the instance of the Red Sea and Mediterranean, appears to me the most satisfactory that the case admits of. But then will not its application extend much farther than this particular case ? Will it not somewhat shake our faith in the precise determinations of strata by specific identity or difference of fossils ? If, in one particular age, two very distinct faunas (different in *every species* as I understand you), could co-exist in neighbouring areas, and one of them nearly identical with the fauna which was at a later time to people *the other* area ; if this could happen once, may it not have happened again and again ? And will not such a discovery seriously damage those fine lines of distinction which Prestwich and others are so fond of drawing among the tertiary formations ? I have really

1856. nothing to send you except comments on your own letters, so I will now say something of that from Salzburg.   I do not wonder that you have been delighted with the Gentians, they are a lovely family of plants.   I am well acquainted with the three kinds you mention ; Fanny and I, in '48, gathered them and seven others in Switzerland and Savoy, and one more on the Apennines, eleven in all. Whether Gentiana Germanica is distinct or not from our English G. Amarella, is a disputed point ; it is difficult to find good distinctive characters, but the Germanica has flowers constantly (as far as I have seen) at least twice as large as those of Amarella ; and that independently of the size of the plant.   Perhaps it is a geographical variety.   It is curious that our common Helix aspersa should be wanting in that country.

The proportion of British species among the land shells, however, appears to be large ; nineteen in twenty-five, you say.   I doubt whether so large a proportion of the plants would be British.   Many of the common plants of those sub-alpine districts are wanting on this side the water.   I am going on with De Candolle's " Botanical Geography," but it takes time to read it, as it is not only dry but exceedingly elaborate, and requires close attention, for which however one is well repaid.

I have drawn up a *résumé* of my observations on the S. Jorge leaves, which will be at your service when you return home.

I have promised to give *another* lecture on coal at Bury ; for as I was only able, the first time to take

in one division of the subject which I had meant to 1856.
embrace, and in fact to give them little more than
the botanical history of coal, I shall try now to give
something about the geographical and geological
relations of the different coal fields, and about the
igneous rocks which interfere with them.

Kingsley has promised a lecture on the Study of
Natural History, in January,and the Arthur Herveys,
with whom he is to stay, have asked me to meet
him ; he is a man I have a great curiosity to see,
or rather to hear. The Arthur Herveys made a
tour this summer in the Silurian region, and seem
to have been charmed with Ludlow, and they
collected fossils, and were interested by seeing
basaltic columns on the Clee Hills.

You will be glad to hear of Edward's appointment
to be Secretary to the Cambridge University
Commission. We are all delighted : the office is
one for which I think he is particularly well-
qualified, the salary is very good, and it has been
given in a very handsome and satisfactory way.
With much love to Mary,

I am ever, very affectionately yours,

C. J. F. BUNBURY.

———

Mildenhall,
November 11th, 1856.

My Dear Katharine,
I believe it is somewhere about a century and-a-half
since I last wrote to you ; the fact is, I have had
nothing very particular to tell, and in so uniform a

1856. life as ours, one day slips by after another, almost unnoticed (like leaves falling from the tree—a simile strongly suggested by what I see before my windows just now) till, on looking back, one is amazed at the accumulation.

My cryptogamic collection has received a large addition since I saw you : when Mr. Eagle's library and other things were sold at Bury after his death, my father bought and made me a present of his whole collection of Mosses and Lichens,* upon which he had bestowed great pains and study. The Mosses are a very rich and valuable set, very nearly complete I believe, as regards the British species ; the result of nearly 50 years study and collecting, for some of the specimens were gathered in 1808 ; rich in authentically named specimens from old Dickson (who is so often mentioned in the English Botany), from Mr. Dawson Turner, Mr. Brown, Sir William Hooker, and some from Mr. Wilson. I daresay Mr. Brown will remember Mr. Eagle's name. Having been a good deal engaged lately with other branches of botany, I have not yet thoroughly studied the collection, but it is a valuable accession to my herbarium. The Lichens, which are in a large cabinet by themselves, are also, I believe, a very complete collection, but I do not understand them so well. I hope to show you the Mosses the next time you come here.

I have been busy with the fossil leaves from Madeira, and studying the characters of recent

---

* This collection I gave, after my Husband's death, by his wish, to the University of Cambridge.—F. J. B.

leaves, trying to satisfy myself whether one could 1856. safely judge of genera and families by leaves *alone* ; as far as I have gone yet, my experience is against it. I have also been working at the catalogue of my Ferns, but have not been able yet to see my way clearly as to a good principle of arrangement for those attractive but perplexing creatures.

Though I spoke of the uniformity of our life here, it has been by no means a dull or solitary one, for we have had (as I believe you have heard) a succession of visitors, and very pleasant ones, both juvenile and grown up ; but no particularly literary or scientific friends, except our unfortunate twice-wrecked friend, Mr. Smith of Orotava, who was very agreeable. After to-morrow we shall be quite alone, and it will seem quite strange to be so.

Fanny sends her love, and pray give mine to your Husband.

Ever your affectionate Brother,
C. J. F. BUNBURY.

Mildenhall,
December 2nd, 1856.

My Dear Mr. Horner,

Fanny received from you this morning a very kind and agreeable letter, for which she thanks you very much, but she is so overworked just now, so overwhelmed with business,—between making arrangements for company in our house, seeing all sorts of poor people, distributing charity for Lady Bunbury, corresponding with Louisa Scott,

1856. who is studying in London, and looking after the schools,—that I have taken up the pen in her stead, and she will write to you in a few days. I am happy to say that, with all this work she is looking very well, and this severe weather appears to agree with her, which I should hardly have anticipated. I too, though I grumble much at the cold, am very well.

It happens curiously enough that I had anticipated your recommendation of Smith's "Theory of Moral Sentiments;" I began it some days ago, and find it pleasant reading: the style easy, clear, and simple, and the illustrations of human nature and manners very agreeable; of his theory I say nothing at present. I think Sir James Mackintosh, while he praises Smith's style and mode of treating his subject does not admit his theory to be sound. In Mackintosh's Dissertation there are many charming passages, especially moral reflections and sketches of character, but in the more theoretical parts I did not think him always clear,—very likely from my want of familiarity with such subjects. Dugald Stewart's Dissertation I began reading when I was last at your house, but stuck fast, I am sorry to say in the middle of the Baconian Philosophy. I hope to try again some day, and to be more persevering.

Fanny and I are reading Dante together in the evenings. By way of lighter reading (though the subject is not a very light, at least not a cheerful one), I have been reading Chesterton's "Revelations of Prison Life,"—ill written, but curious, instructive, and even entertaining.

Did you see in the *Examiner*, a notice of an exam- 1856. ination paper of the Society of Arts, in which the examiner (I do not mean the newspaper so called), coolly places Tennyson alongside of Shakspeare, and requires the candidate to draw a comparison between Hamlet and Maude !

I have engaged to give a lecture at Bury this winter, and am going to give a rehearsal of it *here:* it is to be again on Coal, a second part of that which I gave them last year. As I then treated the botanical part of the subject, so now I shall take it up geographically, and give them some account of the principal coal fields of the world, their extent, comparative productiveness, and something of the geological relations, particularly in reference to the igneous rocks which have played such curious tricks with them ; also I may bring in some *comparative* notices of their fossil plants, as for instance the great similarity between the fossil floras of Europe and North America and the distinctness of those of the Indian and Australian coal fields. I am sorry I have not the requisite knowledge for treating the subject chemically ; though indeed I believe the matter I have in hand will be quite as much as I can get into one lecture.

Pray give my love to all your family party, and believe me ever

Your affectionate Son-in-law,

C. J. F. BUNBURY.

Milton Keynes UK
Ingram Content Group UK Ltd.
UKHW032320161024
449665UK00001B/39